Poverty and the Environment

E N V I R O N M E N T
A N D
D E V E L O P M E N T

A fundamental element of sustainable development is environmental sustainability. Hence, this series was created in 2007 to cover current and emerging issues
in order to promote debate and broaden the understanding of environmental
challenges as integral to achieving equitable and sustained economic growth. The
series will draw on analysis and practical experience from across the World Bank
and from client countries. The manuscripts chosen for publication will be central
to the implementation of the World Bank's Environment Strategy, and relevant to
the development community, policy makers, and academia. In that spirit, this series
will address environmental health, natural resources management, strategic environmental assessment, policy instruments, and environmental institutions.

Also in this series:
International Trade and Climate Change: Economic, Legal, and Institutional
   Perspectives

Poverty
and the
Environment
Understanding Linkages
at the Household Level

THE WORLD BANK
Washington, DC

ISBN: 978-0-8213-7223-4
e-ISBN: 978-0-8213-7224-1
DOI: 10.1596/978-0-8213-7223-4

Cover photos:
Cooking fire: Ray Witlin/World Bank
Misty rainforest: ©Frans Lanting/Corbis

Library of Congress Cataloging-in-Publication Data has been
applied for.

C O N T E N T S

vii   *Acknowledgments*

C H A P T E R   1
**1   Understanding Poverty-Environment Linkages
at the Household Level**
2   Poverty and Environmental Change at the Macro Scale
5   Environmental Management and Pathways to Household Welfare
7   Scope of the Report
9   Some Key Conclusions

C H A P T E R   2
**11   Local Natural Resources, Poverty, and Degradation:
Examining Empirical Regularities**
12   The Importance of Environmental Income to the Poor
15   Commons as a Source of Insurance
18   The Effect of Growth on Local Resource Use
20   Welfare Impacts of Degradation
22   The Role of Poverty in Environmental Change
24   Conclusions

C H A P T E R   3
27   Health Outcomes and Environmental Pathogens
28   Theoretical Linkages between Health Outcomes and Environmental
     Conditions
33   Empirical Evidence of Linkages between Health Outcomes and
     Environmental Conditions
41   How Robust Are the Empirical Findings?
42   Conclusions and Tentative Policy Implications

C H A P T E R   4
45   Household Welfare and Policy Reforms
46   Selected Policy Reforms: Evidence from Case Studies
57   Challenges and Data Limitations
59   Conclusion

C H A P T E R   5
61   Directions for Change
61   Use of Local Natural Resources
63   Design Principles for Improving Environmental Health
63   Better Data for Monitoring Change

64   Policy Reforms for Managing the Environment and
      Reducing Poverty
65   Moving Forward

67   *References*

BOXES

12   **2.1** Poverty and the Environment in Cambodia
17   **2.2** The Role of Natural Resources in Providing Insurance before and
      after Hurricane Mitch
36   **3.1** Does Health Information Increase Households' Efforts to Purify
      Water?
38   **3.2** Diminishing External Benefits from Community Coverage of
      Water and Sanitation
48   **4.1** Impact Evaluation Methods
50   **4.2** Who Participates in the Community Management of Environmental
      Resources?
51   **4.3** Who Benefits Most from Community Management?

FIGURES

4   **1.1** Rank Correlations between Poverty and Various Environmental
      Indicators in Cambodia, Lao PDR, and Vietnam
5   **1.2** Poverty-Environment Linkages at the Household Level
21   **2.1** Biomass Availability in Malawi
22   **2.2** Effect of Increase in Biomass Density on Welfare of Rural Poor in
      Malawi

TABLES

2   **1.1** Selected Macro Indicators Linking Poverty, Natural Resources, and
      Under-Five Mortality
3   **1.2** Health Outcomes and Access to Environmental Infrastructure in
      Selected Countries, by Wealth Quintile
15   **2.1** Environmental Income as Percentage of Total Income in Resource-
      Poor and Resource-Rich Areas
30   **3.1** Factors Affecting Child Health

Acknowledgments

This book is a product of the poverty-environment work program in the Environment Department of the World Bank. The Policy and Economics Team, where this work is housed, is led by Kirk Hamilton.

The book is a collaborative effort, led by Priya Shyamsundar, with contributions by Sushenjit Bandyopadhyay, Limin Wang, Mikko Paunio, and Kirk Hamilton. It draws on the work of many valued colleagues, both inside and outside the Bank, including Patricia Silva, Stefano Pagiola, Agustin Arcenas, Hanan Jacoby, Mei Xie, Bas van der Klaauw, Alessandro Baccini, Keshav Raj Kanel, Irina Klytchnikova, Dragana Radevic, Ana Rios, and Michael Humavindu. The volume also benefited from the comments of James Warren Evans, Laura Tlaiye, Jan Bojö, Kulsum Ahmed, Anjali Acharya, and Muthukumara Mani.

Special thanks are reserved for the manuscript's peer reviewers, Christopher Barrett and Maureen Cropper, who provided invaluable expert advice and guidance on this work. Maureen in particular deserves our thanks for her substantive contributions to the volume.

Finally, the generous financial support of the Government of Sweden for this work is gratefully acknowledged.

C H A P T E R   1

Understanding Poverty-Environment Linkages at the Household Level

THE WORLD BANK'S PRIMARY GOAL IS TO REDUCE POVERTY. It is therefore important to examine the Bank's many lending and development activities through a poverty lens.

In 2002 about half the world's population subsisted on less than $2 a day. About 44 percent of all households in Africa and 31 percent of people in South Asia lived below the $1-a-day poverty line (World Bank 2006d). As these figures suggest, the Bank's poverty mandate remains vast, important, and urgent.

The Bank is one of the largest international donors in the area of environmental management. In 2006 alone, it provided $1.4 billion (in loans or grants) in aid to poor countries to improve the environment. The Bank's activities in this arena include lending for forestry operations, improvements in air quality, changes in environmental institutions and governance, and investments in water and sanitation infrastructure. In partnership with the Global Environment Facility (GEF), the Bank plays a major role in global efforts to stem climate change, biodiversity degradation, and the impact of toxic and chemical waste.

Are these large investments in poverty reduction and environmental management mutually reinforcing? History ultimately will provide an answer; in the meantime, smaller issues can be addressed. An important component of this question, for example, is whether—and to what extent—environmental management can contribute to poverty reduction. Are current environmental management

1

strategies addressing the problems of the poor? What challenges do these strategies face? Most important, what is the role of the poor and what are their behavioral strategies as management programs are put forth?

Poverty reduction is a three-part problem. It involves stemming the fall of households into deeper poverty, enabling poor people to move out of poverty, and preventing the nonpoor from becoming poor. Reducing vulnerability is as important as reducing poverty. While there is a role for environmental management in each of these areas, the importance and type of management will differ.

It is important to take a micro view of the poverty-environment nexus and to understand how households rely on the environment, what factors condition household dependence on the environment, and the extent to which improvements in environmental management change the choices the poor face. These questions are at the core of this report. It focuses on two classes of poverty-related welfare outcomes: income and expenditure measures and health outcomes. The attention to household-level analyses and actions distinguishes this report from other, more broad-based analyses.

Poverty and Environmental Change at the Macro Scale

In order to understand the scale of the poverty-environment problem, it is important to first consider some cross-country indicators of poverty and environmental change (table 1.1). These data indicate that poor countries are much more dependent on natural resources as assets than rich countries. The ratio of people to forested land is more than three times higher in low-income than in high-income countries. This figure gives a crude indication of pressure on forests. While forested lands are growing at 0.1 percent a year in high-income countries, they

TABLE 1.1
Selected Macro Indicators Linking Poverty, Natural Resources, and Under-Five Mortality

Item	Low-income countries	High-income countries
Share of natural resources in total wealth (percent)	29	2
Population per square kilometer of forest	324	104
Deforestation rate (percent per year)	0.5	−0.1
Access to improved water source (percent of population)	75	99
Access to improved sanitation (percent of population)	36	...
Under-five mortality (per 1,000 live births)	122	7

Notes: Wealth-share data are for 2000; all other data are for 2004. '. . .' indicates no data.
Source: World Bank 2006d, e.

are shrinking at 0.5 percent a year in low-income countries. Access to "environmental infrastructure," in the form of improved water and sanitation, shows a similar divide. The outcome is that mortality rates for children under five are nearly 18 times higher in low-income than in high-income countries.

The same general picture emerges from examination of the distribution of health outcomes and access to environmental infrastructure across wealth quintiles within selected developing countries (table 1.2). Wealthier households within these countries have greater access to environmental infrastructure and better health outcomes (less stunting and under-five mortality).

TABLE 1.2
Health Outcomes and Access to Environmental Infrastructure in Selected Countries, by Wealth Quintile

Item	1 (lowest quintile)	2	3	4	5 (highest quintile)
Under-five mortality (per 1,000 live births)					
Egypt	147	119	85	62	39
India	155	153	120	87	54
Kenya	136	130	92	85	61
Peru	110	76	48	44	22
Uzbekistan	70	44	55	52	50
Stunting (percent)					
Egypt	38	34	29	25	20
India	60	59	54	48	34
Kenya	44	38	30	31	17
Peru	46	31	19	10	5
Uzbekistan	40	30	30	25	31
Access to improved water (percent)					
Egypt	47	73	87	97	99
India	6	15	27	44	74
Kenya	1	9	16	43	76
Peru	14	60	87	97	100
Uzbekistan	47	59	78	96	99
Access to improved sanitation (percent)					
Egypt	46	78	94	97	100
India	0	0	4	22	80
Kenya	0	1	3	12	64
Peru	0	7	44	87	100
Uzbekistan	0	1	2	5	70

Note: Data are for 1995 for Egypt, 1992–93 for India, 1998 for Kenya, and 1996 for Peru and Uzbekistan.
Source: Rutstein and Johnson 2004.

The leading health risk factors in developing counties are (in order) malnutrition, unsafe sex, unsafe water and lack of sanitation and hygiene, and indoor smoke from solid fuels (WHO 2002). The prevalence of malnutrition is not only associated with food insecurity, it is now also widely recognized that an unhygienic environment is a key determinant of malnutrition among young children. Clearly, achieving the Millennium Development Goal (MDG) health targets requires public policies that focus on reducing environmental risk factors through better access to basic environmental services, as well as better access to health and education services.

Data from a study conducted in Cambodia, the Lao Peoples Democratic Republic (Lao PDR), and Vietnam by the World Bank's East Asia Region show rank correlations between poverty indicators and environmental indicators (figure 1.1). These data provide evidence of a significant correlation between poverty and certain environmental and health indicators. This macro evidence is not uniform, however, and begs for a more careful examination through micro studies.

These macro indicators suggest that a link between natural resources, the environment, and poverty is at least plausible. Moving the analysis to the household

FIGURE 1.1

Rank Correlations between Poverty and Various Environmental Indicators in Cambodia, Lao PDR, and Vietnam

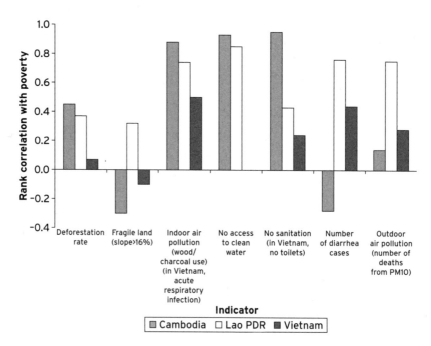

Source: World Bank 2006b.

level allows researchers to examine the correlation further and to identify cases in which the correlation is strong. This is the main focus of this report.

Environmental Management and Pathways to Household Welfare

Environmental change, particularly of local natural resources, can affect poverty through many pathways (Reardon and Vosti 1995; Duraiappah 1998; Wunder 2001; Dasgupta 2003, 2004; Sunderlin and others 2005). To see this relationship more clearly, this section builds on a simple model by Barrett (2004) that links household income and assets.

Consider a poor household whose welfare depends on assets the household has access to or owns. These assets may include biophysical, human, environmental, and constructed capital (figure 1.2) At any point in time, household well-being depends on the returns to these assets and any exogenous shocks (unexpected changes as a result of natural disasters, death, gifts, or macro market changes). Returns to assets generally have two components: known returns and an uncertain component that depends on weather, health, and other factors. Changes in welfare can thus result from four types of changes: changes in asset holdings, changes in returns to these holdings, changes in the

FIGURE 1.2
Poverty-Environment Linkages at the Household Level

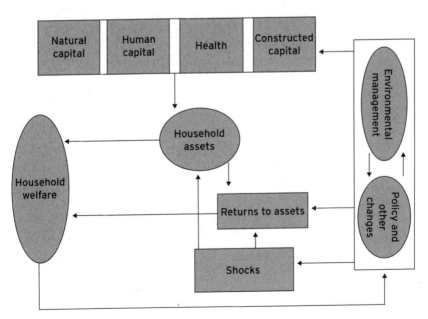

Source: Authors

uncertain component of returns, and changes in exogenous income, which can be positive or negative (Barrett 2004).[1]

Changes in environmental management can have two effects on household welfare in the short to medium term. First, they can change the return to assets. Agro-forestry techniques may improve the productivity of household landholdings, for example; smokeless stove programs may contribute to improved indoor air quality, health, and productivity. Thus, one reason to improve environmental quality would be to add value to the flows from household land or labor. Any health improvements that come from environmental management will also have direct welfare impacts that are independent of productivity improvements.

Changes in resource management can also increase household assets. This could occur, for example, as a result of land reforms or community forestry programs that provide households with secure access to forests. Improved environmental quality may also reduce morbidity or mortality and increase labor power. It is important to recognize that labor is often the only asset that poor households have and that sickness and death can have intergenerational effects. Any improvements in environmental health can have long-term impacts on households' ability to move out of poverty.

Environmental changes can contribute to unexpected shocks over the longer run. Climate change can increase the variability of returns, for example: greater variation in rainfall patterns is likely to increase the variability of crop yields. New disease vectors emerging from climate change may make households more vulnerable. Exogenous shocks, such as floods or hurricanes, can also wipe out household assets and contribute to loss of life. Environmental management matters to the extent that natural barriers such as mangroves and coral reefs diminish the effect of these shocks.

There may also be interactions between poverty and the environment. However, these simultaneous and ongoing changes are difficult to isolate empirically.

This simple model captures the fact that households care about expected welfare outcomes as well as variations in these outcomes. For poor households that are unable to bear shocks, maintaining a steady but low level of economic activity may well be the optimal strategy (Barrett 2004). Such households are simply unwilling or unable to make the changes required to build up their assets or improve their productivity to get themselves out of poverty. Very poor households, for example, may not gain from growth in ecotourism or an increase in the demand for local forest products (Lybert, Barrett, and Narjisse 2002). Even if the returns are high, they may not participate in new jobs, because of perceived risks of switching to new types of labor or because the initial costs associated with switching are too high.

Similarly, access to clean water improves child health, but obtaining a new connection to the main distribution line may be too expensive for the poor (World

Bank 2005). In addition, many households in low-income countries may be uninformed about mechanisms with which to mitigate the effects of poor water quality or the health risks of staying indoors during peak emission periods of cooking with biomass fuels. Better environmental conditions at the community level can generate external health benefits. But public decisions often overlook the health benefits of information and community-level externalities. Many of these issues can converge to keep poor households in low-equilibrium poverty traps.

The existence of poverty traps is particularly relevant for households that are dependent on local natural assets or livestock for subsistence. In such instances, the returns to assets are often endogenous. Among migrant herders in rural Ethiopia, for example, the profits from livestock farming depend on the number of animals in the herd. Lybert and others (2004) find that if an external shock pushes the herd size below a certain threshold, these migratory farmers become sedentary and are no longer able to raise their main asset (livestock). In many cases, fish stocks that have been depleted beyond a certain threshold have not been able to recover. If this happens, the only way out of poverty may be migration and new forms of employment. Even with a small boost, poor households may not be able to pull themselves out of poverty, even in the context of a growing economy.

It is useful to understand the dynamics of poverty and the use of natural commons. The poor are known to decrease short-term consumption in order to maintain the long-term health of their private assets. However, they may also reduce the quality and quantity of the natural capital they have access to in order to increase current consumption, to the detriment of future consumption. As Dasgupta (2004) argues, there can be dynamic feedback loops among poverty, local natural resources, and population growth. Households that depend on the commons may have more children to help them collect from the commons, which can lead to further degradation. In turn, this can trigger a demand for more children. Such action is more likely to occur under conditions of open access or ambiguous tenure over resources. Many of the recent community-based natural resource management programs are an attempt to clarify rights and responsibilities over natural resources in order to minimize such actions. Given high discount rates, however, poverty can lead to depletion of natural capital even when rights are clear. When natural capital is not substituted by other forms of investment, a dynamic spiral can be created in which income and resources decline over time.

Scope of the Report

This report presents micro evidence on how environmental changes affect poor households. It focuses primarily on environmental resources that are outside the private sphere, particularly commonly held and managed resources, such as forests, fisheries, and wildlife.

Three features of the report are noteworthy. First, it uses a data-driven approach to examine the dependence of the poor on natural resources. Considerable case study evidence is available about the poor and their reliance on resources, and various theories have been proposed about the pathways through which changes in the stocks of resources affect the poor. There is an information gap regarding the nature of the dependence of the poor on natural resources and the mechanisms that influence this dependence. This report examines whether household data across large populations and multiple case studies provide evidence of poverty-resource linkages.

Second, the study examines the role of the environment in determining another aspect of poverty: health outcomes. International aid organizations interested in health often focus on building the hardware of institutions and medical supplies or on crafting policy reforms that affect only the health sector. There is a need to broaden the scope of health sector activities to include environmental management as a mechanism for preventing sickness.

Third, the study looks at the role of policy instruments and reforms. Understanding the consequences of policy reform in one sector on outcomes in other sectors is critical, particularly if the changes affect poverty reduction. One topical policy issue is decentralization of natural resource management and the creation of communitarian institutions by the state, partly in response to state-level failures to manage natural resources efficiently. How effective have these institutions been in improving the lot of the poor? Are these institutions egalitarian in their outcomes? This report looks at evidence from multiple countries to address these questions.

Another environmental management tool is payments for environmental services (PES), a mechanism that has been used in a growing number of contexts in Latin America. The report examines the poverty impacts of PES and the willingness of the poor to participate in such schemes.

The report draws on the general economics literature as well as on data collected by the World Bank and its partners to analyze poverty-environment linkages at the household level. The data come mainly from household surveys, such as Living Standards Measurement Surveys. Although the data were not necessarily collected to answer questions about environmental changes and their links to poverty, they include information on a broad range of poverty indicators that can be exploited for this purpose.

Poverty-environment linkages are inherently dynamic and involve behavioral responses that make the identification of cause and effect difficult. Thus, questions related to these linkages are ideally answered with the use of panel data sets or data from randomized experiments. Detailed panel or experimental data are rarely available, however, and there is merit to identifying empirical regularities through rigorous examination of cross-sectional data. This study discusses some of the methodological challenges faced in analyzing poverty-environment problems throughout this report, examining some of these issues in detail in chapter 4. The report also fills important gaps with information drawn from peer-reviewed literature.

Some Key Conclusions

Several conclusions emerge from this study. They can be summarized under four broad headings.

Environmental Income Matters to the Poor

- Natural resources are a significant source of income for many households. They can also provide insurance during times of need.
- In the absence of policy reforms, economic growth is likely to increase resource use in the short to medium term. Both poor and nonpoor households will contribute to resource degradation.
- The high discount rates of the poor and high population growth will likely mean continued degradation of local natural resources.
- The impact on welfare of slow and small changes in resource availability is small, which may encourage resource degradation. As degradation occurs, households use alternate resources or obtain their resources from alternate areas. The low opportunity cost of time in poor households implies that the welfare impact of degradation is likely to be small.
- Poverty reduction will need to be linked to parallel environmental management strategies if the aim is to conserve natural resources or environmental services. Poverty reduction efforts alone will not necessarily increase environmental quality unless specific environmental reforms are undertaken.

Health Outcomes Are Strongly Linked to the Environment in Poor Countries

- Design of health programs and projects should be based on considerations that extend beyond the health sector to include the environment, education, nutrition, and information on public health.
- Public investment in environmental infrastructure should target poor communities rather than poor households, because investment in clean water and sanitation creates positive externalities for household health.
- The role of information has been largely overlooked in many health-related studies, and the role of health information is often ignored in public policy. The lack of public information about the health impacts of poor water quality and exposure to indoor air pollution may reduce the demand for better environmental quality and limit household behavioral responses.
- While it is widely recognized that the use of biomass fuel poses a health risk to poor households, the factors that determine exposure and the types of policy interventions that can reduce exposure require further study. The contributors

to human exposure include energy technology, housing characteristics, and behavioral responses (for example, who does the cooking within households and the amount of time individuals spend indoors during peak cooking periods).

■ There is a strong need to increase the robustness of empirical evidence on environmental health, including through the collection of longitudinal household survey data and the incorporation of questions on cause of death and other retrospective information on social, environmental, and health conditions.

Environment and Natural Resource Reforms Can Improve the Welfare of the Poor

■ Community-based natural resource management yields a measurable improvement in household welfare, stemming from increased economic activity, investment in community infrastructure, and improved management of resources.

■ The extent to which households participate in community-based management of natural resources has mixed impacts on household welfare, with some studies showing that participants derive larger benefits than nonparticipants and others indicating that participants and nonparticipants share benefits equally.

■ Measuring the distribution of benefits from policy reforms can indicate whether vulnerable groups receive the benefits, leading to better targeted reforms in the future.

Better Analytical Tools Are Needed

■ Conducting randomized social experiments and collecting data from before and after policy reforms yield the most-robust analytical results on the impacts of reforms. These research methods are not always practical, however.

■ It is difficult to attribute causality between environmental reforms and poverty alleviation from cross-sectional household data. However, with appropriate treatment and control groups and the use of suitable analytical tools, it is possible to draw policy conclusions from such data.

Note

1 While household income and welfare are used interchangeably here, physical assets are only one measure of well-being. Health is another measure; aspects of this discussion apply equally to health outcomes.

CHAPTER 2

Local Natural Resources, Poverty, and Degradation: Examining Empirical Regularities

RURAL HOUSEHOLDS MAKE UP a large proportion of the world's poor. While markets and infrastructure such as roads, irrigation dams, and water pipes have made their way into their lives, many millions of households still depend largely on two assets for their subsistence: labor and nature's capital. Is this reliance on natural assets significant enough that investments in nature can contribute to poverty reduction?

This chapter addresses three questions related to the dependence of the poor on natural resources:

- To what extent is the environment important to poor households, in terms of both contributing to household income and reducing variations in household consumption?
- As households move out of poverty, is it reasonable to expect their dependence on natural resources to decline?
- What kinds of welfare losses do the poor bear as a result of resource degradation? What are appropriate strategies for poverty reduction and conservation?

Analyzing causal linkages between poverty and natural resource degradation is not easy. The prevalence of feedback loops between natural resource changes and household use of these resources; the reliance of researchers on cross-sectional data because of the lack of good time-series information; and differences in local conditions (markets, resources, infrastructure, customs, and so on) make it hard to reach general conclusions. However, some connections that occur on a regular basis can be observed. This chapter summarizes these linkages based on a review of recent analytical work from within and outside the World Bank.

The Importance of Environmental Income to the Poor

Economic analyses two or three decades ago focused on agricultural farm income, often neglecting the role of other forms of off-farm labor income, petty trade, remittances, and other types of jobs and income that support the rural economy. Researchers now understand that there are multiple sources of income in rural areas and that households often diversify and support themselves with earnings from various sources.

Income that is still frequently neglected, however, is income from natural resources, such as forests, fisheries, and wildlife. National income accounts and estimates of rural household often exclude real income that accrues to households from village commons, state-owned forests, or open-access aquatic resources. This can lead to an underestimation of the use of local natural resources by the poor and an overestimation of poverty (Cavendish 2000; Vedeld and others 2004; Sjaastad and others 2005). Box 2.1 examines the question of resource use in Cambodia.

Of interest to policy makers is how much environmental income contributes to the lives of the poor. The literature on this issue highlights the difference between *use* and *dependence* (Cavendish 2000; Narain, vant Veld, and Gupta 2005; Chettri-Khatri forthcoming). *Resource use* generally refers to the amount of resources consumed or collected by subsistence households; *dependence* refers to the contribution of resources to overall household income. This distinction is important, because resource use and dependence can differ considerably among poor and nonpoor households. Is this dependence worth worrying about in poverty reduction strategies that account for different sources of income obtained by the poor? The issue is difficult to address, because of the multiple definitions of income used in different empirical studies.[1]

BOX 2.1
Poverty and the Environment in Cambodia

Substantial dependence on natural resources exists in Cambodia: nationwide some 72 percent of households collect fuelwood and other forest products, 21 percent collect nonwood forest products, and 53 percent catch fish or seafood (World Bank 2006b). In rural areas, engagement in these activities is more than twice as high among households in the poorest quintile than it is among households in the richest quintile (box figure).

(continued)

BOX FIGURE
Rural Cambodian Households Engaged in Natural-Resource-Dependent Activities, by Consumption Quintiles, 2004

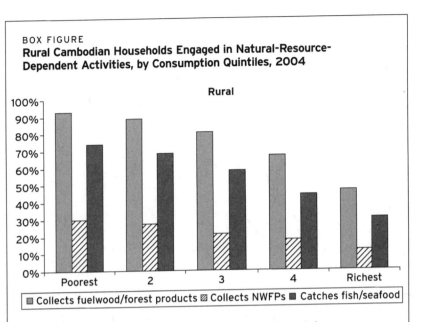

The World Bank study presents some evidence on trends in resource depletion by drawing on an opinion poll of commune leaders. A large percentage of leaders believe that forests and fisheries are on the decline (box table). A much smaller percentage (19 percent) believe that the number of people able to secure livelihoods from natural resources will decline by 2010. The results of this simple survey suggests that there is a low likelihood of resource dependence decreasing in the short to medium term.

BOX TABLE
Responses from Commune Leaders on Natural Resource Decline in Cambodia (percent)

Poll Item	Percentage of Commune Leaders Reporting Decline
Volume of fish catch compared with five years ago	86
Forest cover compared with five years ago	72
Number of people with access to land for cultivation compared with five years ago	28
Number of people projected to be able to secure livelihood from natural resources in 2010	19

Source: Siela and Danida 2005.

The study that originally brought the most attention to the link between environmental income and the poor was N. C. Jodha's (1986) work on village commons in India. Based on data from 82 villages, he found that on average, poor rural households derived 9–26 percent of their income from common-property natural resources, while rich households derived 1–4 percent of their income from this source. Jodha's study suggested that the commons in India, however degraded, were important to the livelihoods of the poor.

Cavendish (2000) studies resource dependence in 29 villages in rural Zimbabwe, measuring income obtained by households from all sources over two periods in an agro-pastoral area that cannot be classified as resource rich. His findings represent one of the best examples of how a careful accounting of local natural resources can improve estimates of poverty and well-being. According to Cavendish, 35–37 percent of the income of rural households came from environmental sources in 1996–97. The richest 20 percent of households obtained about 30 percent of their income from nature; among the poorest 20 percent of households, 44 percent of total income could be considered environmental. Based on his rich data set, Cavendish concludes that "environmental income over and above income sources normally captured by rural household surveys would have boosted measured mean income by as much as 47.3 percent in 1993–94 and 46 percent in 1996–99."

In a micro study of two villages in the forested middle hills of Nepal, Chettri-Khattri (forthcoming) finds wide differences in environmental income (which he defines as income from nontimber forest products) based on the type of property rights held over the commons. In the village with a community management user group, environmental sources contributed some 2 percent of income to the poorest quartile of households and 1 percent of income to the least poor. In the other village, where there was no formal user group but looser informal rules over the commons, 20 percent of the income of the poorest and 14 percent of the income of the least-poor households came from the commons. Chettri-Khattri argues that rules of access are the most important factor in explaining this significant difference in environmental income, although other factors, such as access to markets and employment, may also play roles.

Narain, vant Veld, and Gupta (2005) examine 60 villages in the Jhabua district of Madhya Pradesh, in India. In contrast to Chettri-Khattri's study district, which is almost 60 percent forested, Jhabua is only 19 percent forested; 54 percent of the land is considered agricultural, with the rest is classified as degraded. The study villages were selected to maximize variation in forest stock. The results show that the poorest and least-poor quartiles obtain about 18 percent of their income from the commons, while middle-income groups obtain more. Across all income quartiles, dependence on resources is much lower in resource-scarce areas than in resource-rich areas. These results are similar to those Chettri-Khattri obtained in Nepal.

Can these results, and those of similar studies, be generalized, and if so, to what extent? A meta-study commissioned by the World Bank (Vedeld and others 2004) attempts to answer this question, at least partly. This study examines 54 case studies around the world, 61 percent of them from Africa. The cases reflect a sample of communities in rural areas at the fringes of or within tropical forests. While all meta-studies encounter problems emerging from differences in the underlying case studies, this study nevertheless offers insights into general trends. The authors find that about 22 percent of household income can be attributed to forests. Environmental income contributes more to the incomes of the poor (32 percent) than to the nonpoor (17 percent).

Research suggests that local natural resources contribute to the welfare of the poor, in some cases significantly (table 2.1). This finding cannot be generalized to all rural households, however. It applies mainly to households living on the fringes of forests and households that are largely dependent on natural resources for their subsistence. Some case studies suggest that poor households are dependent on the commons even in areas in which resources are scarce or less accessible; this dependence is lower than in biomass-rich or more accessible areas.

For households that are largely dependent on natural resources, the investments that improve their well-being are not necessarily related to natural resource management. Roads, for example, may allow them to market their forest or agricultural products. Health and education may be their best way out of poverty. Even with investments that are directly related to resource sectors, it is useful to recognize that some investments impose costs on the poor; the net benefits of such investments may not be sufficiently high for local communities to want them.

It is important to ensure that resource-dependent households are not cut off from using natural resources. Large changes in access or availability are likely to have significant effects on the poor.

Commons as a Source of Insurance

Having established that natural resources are a neglected source of household income, the next step is to ask if they have a role in reducing household income

TABLE 2.1
Environmental Income as Percentage of Total Income in Resource-Poor and Resource-Rich Areas

Study	Resource-Rich Areas		Resource-Poor/ Low-Access Areas		Average	
	Poor	Rich	Poor	Rich	Poor	Rich
Jodha (1986)					9–26	1–4
Cavendish (2000)			44	30		
Vedeld and others (2004)[a]					32	17
Narain, vant Veld, and Gupta (2005)	41	23	18	18		
Chettri-Khattri (forthcoming)[b]	20	14	2	1		

Note: In most (but not all) cases, "poor" refers to the poorest 20 percent and "rich" to the richest 20 percent of households. Definitional differences make comparisons across studies very difficult.
a. Data reported are from multiple earlier studies.
b. Nontimber forest product (NTFP) income only.

or consumption risks. Commons—particularly forests, wild animal populations, and fisheries—can act as providers of insurance during times of stress. This role can be very important in developing countries, where financial and insurance markets are thin. It is particularly important in marginal areas within these countries, where social networks may be the sole alternate source of insurance.

Over the years, there has been considerable discussion of the role of natural resources as safety nets or the poor person's bank. There is both conceptual and empirical justification for this idea. Baland and François (2005) develop a theoretical model to show that in situations of asymmetric information or unenforceability of contracts, privatization of the commons can reduce welfare, even if privatization is costless and equitable. Welfare is reduced because of the insurance role of the commons, which is not picked up by private insurance providers.

The empirical literature on the insurance role of natural resources is thin. In general, more is understood about household management of ex ante known risks than about responses to unexpected shocks.

Several studies from Latin America provide initial evidence of the role of resources as providers of natural insurance. Pattanayak and Sills (2001) examine forest collection trips in the Tapajos region in the Amazon in Brazil to determine whether households respond to known agricultural risks and sudden agricultural shocks by increasing their dependence on natural resources. They find that forest-product collection is correlated with agricultural yield risks (income-smoothing response) and unexpected production shocks (consumption-smoothing response). The statistical link between forest trips and known risks is strongly significant and more robust than the link between forest trips and unexpected shocks.

Takasaki, Barham, and Coomes (2004) examine coping strategies in response to covariate flood shocks around Peru's Pacaya-Samiria National Reserve. They find that for the very poor, who have only labor available and few land assets, nontimber forest resources act as a source of insurance during difficult times.

McSweeney (2005) examines the natural insurance provided by forests in Honduras before and after Hurricane Mitch. She uses both quantitative and qualitative approaches to show that forests play a critical insurance role (box 2.2).

These findings are broadly corroborated by work in Asia on the effect of the regional economic crisis in the 1990s on forests. In Indonesia, for example, rural households compensate for the loss of agricultural income by increasing forest use.

Ignoring the buffering function of natural resources can undermine the implementation of environmental management or poverty reduction programs. An example from Africa illustrates this point. Many conservation programs in Africa use the distribution of game meat as a strategy for local conservation and development. Do such programs work once the insurance role of resources are recognized? Barrett and Arcese (1998) use a simulation exercise to examine this question in the

BOX 2.2
The Role of Natural Resources in Providing Insurance before and after Hurricane Mitch

Hurricane Mitch struck the Tawahka community in northeastern Honduras in October 1998. It brought down homes, destroyed virtually all agricultural production, blocked waterways, and contaminated water sources.

Traditionally, the Tawahka economy was based on shifting cultivation, permanent polyculture of riparian plots, and extraction of forest products. Initially after Mitch, forest-product use—wood for house construction, wild foods, and medicinal plants—increased. There is evidence that access to these forest products helped the Tawahka cope with the disaster. Soon after Mitch, however, the government presence in the area and increased monitoring and surveillance led to a decrease in the use of timber for commercial purposes. The decline represented a big adjustment for the Tawahka, who had a long history of trading forest products, particularly canoes.

The community found other ways to cope. Households, especially those that were able and young, increased their claim over upland as the land available for agricultural crops declined. They also looked to wage labor and remittances, despite their deep aversion to migration.

McSweeney concludes that forests provide natural insurance, in terms of land and products, making it feasible for households to recoup after natural disasters.

BOX TABLE
Average Household Assets among the Tawahka before and after Hurricane Mitch

Variable	1998	2001
Number of male workers	1.7	1.4
Number of herd of cattle	1.3	1.3
Share of income from forest products (percent)	15.4	7.8
Hectares of primary forest claimed	1.5	12.3
Share of land in primary forests (percent)	3.8	34.4
Number of cultivated cacao trees	413	97
Number of cultivated peach palm trees	20.2	8.9

Source: McSweeney 2005.

context of the Serengeti ecosystem, where wildebeest meat is distributed to reduce poaching. They find that while this strategy works during normal times, it is likely to fail when natural shocks occur. Wildlife poaching is an important source of sustenance when rains fail and agricultural production declines. During such periods, however, the wildebeest herd is least able to withstand increased harvest. Thus, double pressure is exerted on the wildebeest population, which can lead to a collapse in the population and management strategy.

How much of a buffer do local natural resources provide during difficult times? The answer depends on local conditions and household behavioral responses. Alternate policy prescriptions for dealing with risk (such as insurance schemes) differ from those that deal with shocks (food support or employment schemes). More needs to be known about whether natural resources serve the same type of function in both cases.

Until clear alternate options are available, it makes sense to manage local natural resources as part of a portfolio of assets required to minimize consumption risks and help the poor cope with shocks. Attempts to reduce vulnerability need to pay attention to the role of local natural resources in buffering the poor against market, policy, and environmental uncertainties.

The Effect of Growth on Local Resource Use

It is sometimes assumed that as households and economies become wealthier, they reduce their dependence on local natural resources, reducing pressure on these resources. Is this actually the case?

Theoretically, there are multiple possible outcomes of local economic growth on resource use. Demand for energy, fodder, and water is likely to increase, leading to greater use of these resources and perhaps further resource degradation. Increases in wealth also improve education and awareness and increase the opportunity cost of time, which may reduce the collection of natural resources. Improved awareness may also contribute directly to more discriminate use of resources. For example, households may try to find substitutes for fuelwood in order to reduce indoor air pollution.[2] Economic growth also brings exit opportunities for labor (migration), with consequent reductions in resource dependence.

Local economic growth does not affect everyone in the area evenly. Even if markets open up new opportunities, only part of the local population may be able to take advantage of these opportunities; others may remain as dependent on natural resources as ever. The opening of markets for specific natural resources without proper regulatory systems in place may lead to indiscriminate use. In short, the overall effect of increased development may well follow a Kuznets curve, with large reductions in use appearing only at relatively high levels of income.

What do empirical studies show? The best way to study how growth affects use and dependence is by to examine households over time. However, few studies have access to time-series data. A popular alternate approach is to examine a cross-section of households. Cross-sectional estimates are reasonably good proxies for what may happen over time. They overestimate the impacts of growth, however, because households make adjustments over time (Baland and others 2006).

Several studies provide insights into the empirical relation between increases in income and wealth and resource use. Vedeld and others (2004) use a meta data set to examine the link between environmental income and total income. They find that the income elasticity of environmental income is close to one (that is, a 1 percent increase in total income increases environmental income by 1 percent). Thus, across their sampled rural communities, an increase in total income is closely correlated with a proportional increase in the use of forests and wild products. Though somewhat less robust, a second result is that forest dependence increases and then decreases with total income. The authors break their data set into five income quintiles. Like Narain, vant Veld, and Gupta 2005, they find a bell-shaped relation between income and dependence, with groups in the middle-income categories most dependent on forests. This result is not entirely surprising, because middle-income households with land or livestock are most dependent on forests for complementary goods.

Two World Bank studies—one on India (Bandyopadhyay and Shyamsundar 2004), a second on Nepal (Bandyopadhyay and Shyamsundar 2005)—provide additional insights. Both studies use large data sets on rural households and examine (among other things) the relation between wealth, measured as an index of household durable goods, and fuelwood use. The India study examines factors that affect fuelwood consumption in five states. The Nepal study focuses on fuelwood collection by rural households. The authors find that in India, fuelwood consumption decreases with wealth, while in Nepal it increases. These results suggest that given the availability of substitutes, households move away from fuelwood as a source of energy. Where markets for fuelwood are thin and few affordable substitutes are available, as is the case in Nepal, fuelwood use increases as wealth increases.

Baland and others (2006) corroborate some of these results. Their study is based on data on more than 3,000 households in 161 villages in two states in India in the mid-Himalayas (Himachal and Uttaranchal). The authors find that per capita fuelwood consumption increases with income and decreases with the time costs of collecting wood. The net effect of a simultaneous increase in income and labor costs makes fuelwood demand inelastic with respect to growth. However, population growth is likely to lead to further extraction. The authors argue that based on reasonable assumptions about the future, it is very unlikely that fuelwood extraction will decline in these states without some strong policy measures. Growth by itself is insufficient to stem forest resource use.

Based on empirical evidence, it is difficult to predict whether resource use will decrease or increase as rural areas develop and households are pulled out of poverty. Much will depend on the other factors that prevail. At least given current levels of income and poverty, it may be appropriate to assume that local economic growth in conjunction with a growing population is likely to contribute to more rather than less local resource use.

Welfare Impacts of Degradation

Bandyopadhyay, Shyamsundar, and Baccini (2006) examine the effect of changes in resource availability on poverty in Malawi, where some 95 percent of households use biomass as their sole source of energy. Over the years, this and other factors have contributed to a significant loss in forest cover, particularly in the south and central regions. The situation is so grave that Malawi can be considered a country in biomass distress (figure 2.1).

Given the extreme dependence of households on biomass in Malawi, it is compelling to assume that biomass loss in forests hurts the poor. Using a combination of remote sensing and econometric techniques, the authors study whether this is indeed the case. Their study controls for different types of capital that may influence household welfare and asks if natural capital (forests) has a strong effect on household consumption. They find that 80 percent of poor households are affected by forest scarcity, although the impact is very small, with a 10 percent decrease in biomass availability per hectare reducing consumption of poor households by just 0.1 percent. These results are consistent with those of Baland and others (2006), who find that degradation causes only a very small (less than 1 percent) decline in household consumption in India. Degradation thus continues, because households do not feel the pinch of the local externality they create when they degrade environmental resources.

As biomass increases, welfare first increases and then falls (figure 2.2). The average poor rural household benefits from an increase in biomass stock until it reaches 39 cubic meters per hectare. Eighty percent of poor rural households in Malawi live in areas with less than 39 cubic meters of biomass per hectare. Most of the rural poor would thus benefit if average biomass per hectare almost doubled from current levels of about 20 cubic meters per hectare.

The results of Bandyopadhyay, Shyamsundar, and Baccini (2006) on Malawi and Baland and others (2006) on India suggest that households continue to degrade forests because doing so has very little effect on their own welfare. Others studies find that natural resources contribute significantly to household income. How can these two sets of results be reconciled?

First, the study on India focuses on only one aspect of environmental income—fuelwood—while some of the studies on environmental income account for many different sources of environmental income. The Malawi study looks across a range of rural households, not just households living on forest fringes. Studies such as Cavendish (2000) focus on the average contribution of forests to rural income, while more recent studies look at the marginal contribution. Thus, at the margin, degradation has a very small impact on household welfare. This is not to say, however, that households will not be significantly affected if large chunks of forests are removed.

FIGURE 2.1
Biomass Availability in Malawi

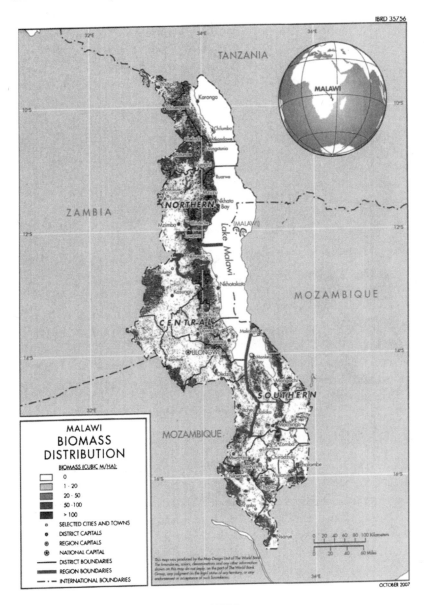

Source: Bandyopadhyay, Shyamsundar, and Baccini 2006.

FIGURE 2.2
Effect of Increase in Biomass Density on Welfare of Rural Poor in Malawi

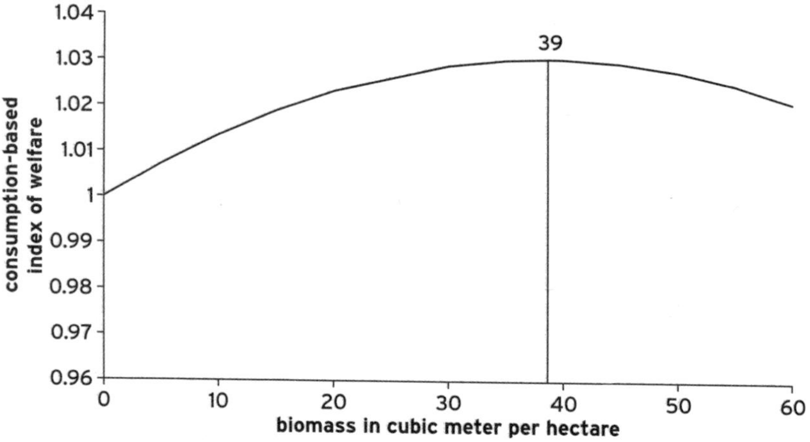

Source: Bandyopadhyay, Shyamsundar, and Baccini 2006.

Moreover, households tend to smooth consumption over space and time. They save (or share) and repay debt during good times and borrow during bad times. Households adjust to slow changes in resource availability, reducing the effect on welfare. They can accommodate small changes over time. In contrast, sudden and chunky changes in forest cover are likely to result in significant declines in welfare.

The Role of Poverty in Environmental Change

Are the poor victims or perpetrators of environmental change? The relation between the poor and natural resources is mediated by a number of micro and macro factors, such as labor and credit markets, property rights, and information about best practices (Bluffstone 1995; Duraiappah 1998; Wunder 2001; Adhikari 2005; Fisher, Shively, and Buccola 2005). Under certain circumstances, it may be optimal for poor and nonpoor people to mine natural resources.

The study by Baland and others (2006) offers some insights into the household subsistence use of fuel and fodder in India. They find that timber accounts for biomass removal of only 48 tons per village, while firewood accounts for 456 tons. Most of the forested areas in this region are degraded; use of the forests for fuelwood appears to be the main cause. Takasaki, Barham, and Coomes (2004) tell a similar story about households living around Peru's Pacaya-Samiria National Reserve. In this area, there was very little timber logging or colonization—less than 1 percent of the area had been transformed into agriculture. Degradation was largely the result of the subsistence use of resources.

Degradation is, of course, only part of the story of environmental loss. In many other parts of the world, commercial timber logging, forest conversion to agriculture, and coral mining have led to major changes in ecosystems (Sunderlin and others 2005). Shifts in relative prices, technological changes, or public investments can trigger such transformations (Kaimowitz and others 1998; Wunder 2001). The plight of the forests, for instance, may depend on whether market, technological, and policy changes favor expanding agricultural land, mechanization of agriculture, or increased urbanization and out-migration of labor.

As a general rule, changes in prices and technology that favor capital-intensive agriculture can contribute to deforestation (Angelsen and Kaimowitz 2001). Where commercial interests are involved, the returns to land-use changes are more likely to accrue to the rich. A case in point is made in the World Bank report *At Loggerheads*, which discusses the intense deforestation that occurred in Brazil between 1999 and 2002. Much of this deforestation can be attributed to increased profits in land uses such as soybean cultivation and pasture development, which were driven by exogenous changes in global markets for soybeans, exchange rates, and other factors (World Bank 2006a). Most of the gains went to large farmers and the wealthy rather than the poor.

Under certain circumstances, poverty may force households to consume assets that may support a longer-term income stream. Silva (2005b) explores this issue econometrically in the context of protected marine areas off the coasts of Tanzania and Zanzibar. The coastal areas of mainland Tanzania are home to 25 percent of its population, while some 1 million people live on the island of Zanzibar. A large proportion of these people depend on fisheries for food and income.

Silva examines the role of destructive fishing gear—such as gillnets, beach seine nets, and drive nets—and practices such as spear gun fishing, poison fishing, and dynamite fishing. She finds that poverty is associated with an increase in the use of illegal gear and practices that are harmful to the marine ecosystem. Female-headed households and households that are food insecure are more likely to use such gear; wealth and education decrease their use. Household welfare, measured in terms of consumption expenditure, increases as a result of the use of destructive gear. This study seems to provide evidence that poverty and environmental degradation can be linked in a downward spiral. However, this is a static representation of a dynamic problem. Whether the poor pull themselves out of poverty as a result of a consumption boost or other factors or whether a consumption boost creates a poverty trap is hard to establish without a better understanding of the dynamics of poverty and natural resource use.

Banning destructive gear, which would be good for the long-term health of the fishery, is likely to hurt the poor. In Tanzania the solution appears to be the creation of alternative income-generation activities. Silva finds that an ongoing alternative income program had a significant impact on household welfare. Thus, a dual approach that imposes costs on the use of destructive gear and provides alternate strategies for increasing income appears to be the way forward. However, such alternative strategies are often difficult to implement.

An interesting twist to the poverty-environmental degradation story in Silva's study concerns the role of credit and ownership of fishing gear. Both factors, which are frequently used as instruments to pull people out of poverty, increase the probability of using destructive gear. Thus, some mechanisms used to reduce poverty may exacerbate environmental problems. Any attempts to reduce poverty through instruments such as credit must be accompanied by strategies to control destructive environmental actions.

Reforming one market can exacerbate a failure in another market. It is therefore very important to recognize the sectoral effects of reforms undertaken for different purposes by different agencies.

Conclusions

Several insights emerge from this review of the evidence on natural resources and household welfare:

- Natural resources serve as a significant source of income to some households. Resources can also serve as a buffer, or insurance, during times of need. Understanding of the insurance role of natural resources is limited.

- Economic growth is not likely to dramatically reduce local resource use in rural areas. In fact, unless significant policy measures are put in place, resource use is likely to grow in the short to medium term.

- Both poor and nonpoor households contribute to resource degradation. In some cases the lack of markets contributes to degradation; in other cases growth in markets can lead to the same outcome. Weak governance institutions, high discount rates, and population growth will all likely continue to contribute to degradation of local natural resources.

- One of the reasons why households degrade natural resources is that the impact of slow and small changes in resource availability on welfare is small. Households adapt to changes over time by using alternate resources or obtaining their resources from alternate areas. As long as the opportunity cost of time is low, the welfare impact of degradation is likely to be small.

- Attempts to reduce poverty will need to be matched with separate environmental management strategies if the goal is to conserve natural resources or services while reducing poverty. Poverty reduction will not necessarily lead to an improved environment unless specific environmental action is taken.

What policy insights does this review provide for environmental management for poverty reduction? Policy makers need to worry about pushing people out of poverty as well as stemming their fall into it. Ensuring that resource-dependent communities have sustainable sources of income from nature is one way to prevent households from experiencing deeper poverty. Discrete and substantive changes in resource availability or access will push the poor deep into poverty, unless these changes directly result in alternate sources of income.

While some evidence suggests that degradation has a smaller effect on household welfare because of how households adapt, this is not its only impact on the poor. Households may feel compelled to make less risky decisions as resource availability becomes less secure. One way to help households make high-risk/high-return decisions, thereby enabling movements out of poverty, is to ensure that resource-dependent households feel they can rely on nature's bank.

The range of choices available to the poor needs to increase. Strategies or technologies that increase the productivity of natural assets would help the poor. Agro-forestry, value addition through the commercial sale of nontimber forest products, and improved local management are some examples of such strategies. (Other such strategies are examined in chapters 4 and 5 of this report.) There are costs to improved resource management, however. Unrealistic expectations about the poverty impacts often stem from ignoring these costs. Of particular importance is the fact that these costs can add additional burdens on local communities and governments, which can contribute to program failure.

Many of the pathways out of poverty will be created outside the natural resources sectors. The most promising investments for poverty reduction may lie in strengthening human capital, including health, and providing infrastructure that allows the poor to access external markets and information (see figure 1.2). From the perspective of poverty reduction, natural assets are not necessarily the assets that will provide the largest payoffs. However, they are assets that cannot be ignored in any investment strategy for poverty reduction.

Notes

1 In its most fundamental sense, environmental income is rent acquired from nature's provision of goods and services. Rent, however, is difficult to estimate. A working definition given by Sjaastad and others (2005) identifies environmental income as rent or value added from "alienation or consumption of natural capital in the first link in a market chain, starting from the point at which the natural capital is extracted or appropriated." This definition is useful, because it limits the scope of products that can contribute to environmental income.

2 Awareness can also contribute to behavioral changes—hand washing, avoidance of pesticide exposure, and so on—that improve health.

CHAPTER 3

Health Outcomes and Environmental Pathogens

IMPROVING HEALTH OUTCOMES OF POOR PEOPLE—by reducing environmental risk factors and increasing access to health and education services— is widely recognized as essential to achieving the MDG health targets. Evidence shows that environmental risk factors account for about one-fifth of the total burden of disease in low-income countries (World Bank 2001). Among the leading health risk factors in developing counties, malnutrition ranks first, unsafe sex second, unsafe water and lack of sanitation and hygiene third, and indoor smoke from solid fuels fourth (WHO 2002). The prevalence of malnutrition, particularly among young children, is also closely associated with environmental risk factors; this is becoming a major focus of World Bank work on malnutrition (World Bank 2006c). Recent studies confirm that factors other than food insecurity and poor child care—such as maternal malnutrition, an unhealthy environment, and poor health care—are also important determinants of the prevalence of malnutrition. In addition, underweight children have a higher risk of mortality from infectious illnesses such as diarrhea and pneumonia, which are caused mainly by exposure to contaminated environmental conditions (WHO 2004).

This chapter reviews the findings of several studies that analyze the links between health outcomes and environmental conditions using household survey data. Environmental conditions, at either the household or community level, are typically defined narrowly, because of data limitations in household surveys.

Key environmental factors include access to water (water sources, types of ownership, and distance to residence); access to sanitation services and disposal of human waste; and access to energy sources (types of cooking fuels). Health outcome indicators used in these studies include child mortality and the prevalence of diarrhea; the prevalence of malnutrition (underweight and stunting); and the incidence of acute respiratory illness among children and adults.

Empirically analyzing environmental health linkages is challenging, because of intrinsic difficulties in measuring health outcomes and environmental quality (as reflected in water quality and quantity and bacterial counts, for example) and because of the complexity of the transmission channels from environmental conditions to health outcomes. In addition, households' behavioral responses affect both health outcomes and access to environmental services. Investment in environmental infrastructure at the household level is likely to also benefit neighbors (that is, private investment may yield external health benefits). Consequently, conflicting findings often emerge from these studies, even when conducted using similar analytical methods and data sources.

This chapter addresses three questions, through a review of empirical studies on the linkages between health and environmental conditions: What are the main analytical issues in environmental health? What are the main findings? How robust is this body of evidence? Based on the findings, the chapter provides some tentative policy recommendations.

Theoretical Linkages between Health Outcomes and Environmental Conditions

Social science, medical, and epidemiological studies of the determinants of health outcomes often adopt different approaches and methodologies (Mosley and Chen 1984). Social science studies focus primarily on statistical associations between socioeconomic and environmental factors and health outcomes—for example, child survival outcomes or nutritional status—using household-level survey data. These studies often do not address medical causes of child death or explain the mechanisms by which socioeconomic (as well as environmental) conditions operate to produce the observed mortality outcomes. Medical research focuses on the biological processes of diseases, attributing mortality to specific disease processes (such as infections or malnutrition) based on death reports collected from clinical sources or recalls from household surveys. Epidemiological studies emphasize the mechanisms of disease transmission in the environment, linking health outcomes with environmental contamination (for example, drinking water, waste disposal, or indoor air pollution). Nutrition studies focus mainly on linkages between breastfeeding, dietary practices, food availability, and nutritional status.

The critical problem with these disparate research approaches is that the selection of a particular research methodology often results in policy and program recommendations biased in favor of a specific discipline. Studies on child malnutrition, for example, often lead to advocacy of particular health interventions, such as feeding programs, largely overlooking the evidence that malnutrition is as dependent on maternal health factors and environmental conditions (poor hygiene as a result of unsafe water and sanitation-induced diarrheal diseases and infections) as it is on nutrient deficiency (Cole and Parkin 1977; Mata 1978; Pinstrup-Andersen, Pelletier, and Alderman 1995).

To address this problem, Mosley and Chen (1984) propose a general analytical framework that incorporates social, economic, and medical science methodologies to study the determinants of child survival. Wolpin (1997) constructed an analytical structural model in the setting of optimal household decision making and identified key issues and associated difficulties in empirical implementation. This framework has provided the base for many empirical studies that have used household surveys as the principal data source for analyzing the determinants of health.

The Analytical Framework

Different factors determine child health during gestation, birth, the perinatal period, the postnatal period, and early childhood (table 3.1). Because the effects of many factors vary with the age of the child, a model of child health must analyze environmental determinants by age.

The factors affecting child health can be divided into factors that affect the likelihood of a child becoming ill or malnourished and the factors that affect the probability of a child dying conditional on becoming ill or malnourished. These factors can then be grouped into nutrition, biological conditions, environmental conditions, and access to heath services. To a large extent, access to services (health care, use of oral rehydration therapy) and environmental conditions (connection to piped water and water quality) are determined by a household's health information; the level of education (in particular of the female head); hygienic behavior (hand washing or water disinfection); intrahousehold resource allocation in food consumption; and other socioeconomic factors (including income). Many factors that affect environmental conditions also affect health outcomes, but they are not directly observable (or difficulty to quantify) in survey data. This data deficiency poses the main challenge in analyzing environmental health linkages using household surveys.

This chapter focuses on studies that examine the impact of exposure to environmental contaminants on child health outcomes. Exposure to diarrheal disease through the oral-fecal contamination route depends on household sanitation (how the household disposes of fecal matter); the availability of water for personal and domestic hygiene; and the quality of drinking water. However, the impact of access to safe water and improved sanitation on exposure depends on the

TABLE 3.1
Factors Affecting Child Health

Factor	Birth and Perinatal	Postnatal	Early Childhood
Factors Affecting Morbidity/Malnutrition			
Nutrition	Maternal nutrition Breastfeeding	Breastfeeding	Food consumption
Biological	Maternal age Maternal disease history Birth order Birth spacing	Birthweight Innate frailty	Innate frailty
Exposure to environmental contaminants	Maternal exposure to indoor air pollution	Access to water (water availability and quality of water) Access to sanitation facilities Cooking fuels and household ventilation conditions	Access to water (water availability and quality of water) Access to sanitation facilities Community-level basic environmental services Cooking fuels and household ventilation conditions
Access to health services	Birth place (clinic versus home, for example) Antenatal care	Immunizations	Immunizations
Factors Affecting Probability of Death, Conditional on Morbidity and Malnutrition			
Treatment	Access to health services	Access to basic medicine (for example, oral rehydration therapy)	Access to basic medicine (for example, oral rehydration therapy)
Nutrition status	Child's nutritional status	Child's nutritional status	Child's nutritional status
Exposure to environmental contaminants		Access to water (water availability and quality of water) Cooking fuels and household ventilation conditions	Access to water (water availability and quality of water) Community-level basic environmental services Cooking fuels and household ventilation conditions

Source: Author compilation.

knowledge and use of good hygiene practices. Exposure to indoor air pollution may increase the incidence of acute respiratory illness. The effect the type of cooking fuel has on exposure to indoor air pollution depends on how much time is spent indoors during periods of peak exposure, which is likely to be influenced by knowledge of health effects.

Two important implications emerge from table 3.1. First, looking at the impact of water and sanitation or cooking fuel on child health should be conditional on parental knowledge of hygiene practices or factors, which may mediate the effects of burning biomass on a child's exposure to indoor air pollution. Second, in measuring the impact of these environmental factors on mortality or morbidity, researchers should control for the other determinants of health listed in table 3.1

Problems Affecting Empirical Research

Studies that use household surveys to analyze the health impacts of environmental conditions aim to obtain unbiased estimates of these impacts with high precision. Four problems commonly plague such studies. One is the inability to control for some of the factors affecting health listed in table 3.1. If these factors are correlated with environmental conditions, estimates of environmental effects will be biased. A second problem is obtaining a sufficiently large sample to detect an effect. This is particularly problematic in studies of infant and child mortality. A third problem is sample selection bias in morbidity studies. If the weakest members of the population have died, the impacts of an environmental condition on a randomly chosen member of the population will likely be underestimated. A fourth problem is errors in measuring environmental conditions, which will likely bias estimates of their effects toward zero.

The bias caused by omitted variables sometimes occurs because household data sets do not contain information on child health status, family access to health care, or parental knowledge of health effects. In some cases, even proxies for these variables—such as family income or assets or maternal education—are unavailable. Omitting these variables is likely to bias estimates of the impact of access to improved sanitation or clean fuels, because these environmental variables are likely to be positively correlated with unobserved factors that improve child health.

One way of handling this problem is to conduct a randomized trial of interventions to reduce exposure to environmental contaminants. The advantage of a randomized trial is clear: if the distribution of an intervention is truly random, it will be independent of other factors—observed and unobserved—that affect health. Randomized trails have been conducted for home drinking water disinfection (Semenza and others 1998; Quick and others 1999, 2002) and for hand washing (Luby and others 2005; Cairncross and Valdemanis 2006). Smith (2006) conducted a randomized trial of improved stoves in Guatemala. Other water and sanitation interventions—such as piped water connections and toilets—are not as amenable to randomized trials. For this reason, controlled experiments are unlikely to be a significant source of data in the water and sanitation area for many years.

Econometric techniques can sometimes be used to deal with the problem of omitted variable bias in observational studies. Propensity-score matching selects households without access to water and sanitation or clean fuels that are observationally equivalent to households with access (see, for example, Jalan and Ravallion 2003). The logic is that households that look similar in terms of their observed characteristics are likely to be similar in their unobserved attributes. If observations on the health impact of (for example) indoor air pollution exist for several household members, household fixed effects (a household dummy variable) can control for unobserved variables common to all household members (Pitt, Rosenzweig, and Hassan 2006). If panel data are available, a dummy variable can be included for each household member to control for unobserved factors affecting health that change slowly over time.

The second problem mentioned above—having a large enough sample to detect an effect when the health outcome is the infant or child mortality rate—may argue in favor of using household surveys. Conducting a randomized trial of sufficient power to detect an effect of an environmental intervention on infant or child mortality would be prohibitively expensive.[1]

The third problem—the sample selection bias resulting from analyzing the health impact of the environment infrastructure using only the surviving population—has long been recognized in the health literature. Children who survive differ systematically from those who die, particularly in high-mortality populations, such as the many African countries in which under-five child mortality rates exceed 100 per 1,000 births. Inferences about the health benefits of infrastructure programs—such as public investment to provide universal access to safe drinking water or sanitation services—can substantially underestimate the effectiveness of health benefits (or even lead to spurious associations) as a result of the failure to take account of the potential reduction in mortality of members of the birth cohorts that died. Most studies of nutritional status that are based only on surviving children are likely to be subject to such sample selection bias if appropriate estimation methods for correcting the sample bias are not applied.

Finally, measuring an environmental exposure with error—using the type of cooking fuel as a proxy for a child's personal exposure to particulate matter, for example—will result in a classic error-in-variables problem. This problem will bias the coefficients toward zero.

Despite these empirical difficulties, many studies analyze the determinants of health outcomes using cross-sectional household survey data. Fewtrell and Colford (2004) argue that these cross-sectional household survey studies are needed to fill the serious gaps in knowledge regarding the effectiveness of sanitation interventions in particular.

Empirical Evidence of Linkages between Health Outcomes and Environmental Conditions

This section summarizes the findings of several important studies, examining the results on four dimensions of health outcomes: child mortality, child morbidity from diarrhea, child malnutrition, and health risks caused by indoor air pollution.

Child Mortality

A large body of literature on the determinants of child mortality has been published in biomedical, demography, and economics journals. Relatively few studies have been conducted that use household surveys to focus primarily on environmental determinants of child mortality. These include studies of Bangladesh and the Philippines (Lee, Rosenzweig, and Pitt 1997); Brazil (Merrick 1985); China (Jacoby and Wang 2004); Ghana (Lavy and others (1996); India (Hughes, Lvovsky, and Dunleavy 2001 and Van der Klaauw and Wang 2005); Malaysia (Ridder and Tunali 1999); and Uttar Pradesh (India) (Bhargrava 2003).

Access to safe water. The studies on China (Jacoby and Wang 2004) and India (Hughes, Lvovsky, and Dunleavy 2001, using 1992–93 National Family Health Survey Data [NFHS], and Van der Klaauw and Wang 2005, using 1998–99 NFHS data) use hazard functions to estimate the impact of environmental factors on child mortality risk. One of the major advantages of these studies is their large sample size. The total number of live births was 160,899 for the China study, 59,000 for India in 1992–93, and 53,201 for India in 1998–99. A large sample size of household survey data, with a sufficient number of observations on child deaths, is particularly important for obtaining a statistically significant estimate of the impact of household as well as community access to environmental services on child mortality.

Improving access to safe water sources is generally regarded as one of the most critical preventive environmental measures for reducing child mortality and morbidity. However, empirical studies based on household surveys do not provide consistent evidence to support such a premise.

The study on China by Jacoby and Wang (2004) provides strong evidence that access to safe water sources is associated with lower child mortality rates. The estimates show that the largest and most significant impact on child mortality comes from access to safe water, which includes water sources from pipes, inside household or public taps, and deep wells within a short talking distance. The results estimate that improving access to safe water from the average level of 33 percent in the early 1990s to universal access in rural China could reduce the under-five child mortality rate by 9 percent (from 33 to about 30 deaths per 1,000 births). Targeted investments can generate a significantly larger health impact: improving safe water access to poor localities increases the health benefit by about 33 percent (in terms of mortality rate reduction) over untargeted investments. These results on the child mortality benefits of access to safe water are emphasized in work by Cairncross and Valdemanis (2006) on disease control priorities in developing countries.

Using information collected on causes of death in the China health survey, Jacoby and Wang (2004) also attempt to validate the causal interpretations of access to safe water on reductions in child mortality. Their results show that the probability of child death from causes that should not be associated with safe water (birth-related deaths and neonatal tetanus) is indeed unrelated to access to safe water. While the probability of dying from diarrheal diseases is, as expected, most responsive to interventions that improve access to safe water, the China study also provides a statistically significant association between safe water access and fever/acute respiratory illness. This emerging and potentially very important water hygiene–infectious agent transmission pathway for acute lower respiratory infections, which kill 2 million children a year, is not yet recognized to be part of the water, sanitation, and hygiene risk factor in the WHO global burden of disease estimates, but it appears strongly in a randomized trial in Karachi reported by Luby and others (2005).

Studies on India using 1992–93 NFHS and 1998–99 NFHS data find no significant impact of household-level access to safe water on the probability of child survival.[2] Hughes, Lvovsky, and Dunleavy (2001) show that improving community access (that is, increasing coverage within a community) to safe water or sanitation in both urban and rural areas significantly reduces child mortality risks. The studies by Lee, Rosenzweig, and Pitt (1997) on Bangladesh and the Philippines and by Ridder and Tunali (1999) on Malaysia provide no evidence of health benefits of access to safe water sources. The relatively small sample size in these studies (611 for Bangladesh, 837 for the Philippines) may be the reason for the lack of statistical significance of the results.

Access to electricity. Several studies find a statistically significant impact of access to electricity on child mortality. The India studies using NFHS data indicate that access to regular electricity supply significantly improves children's survival chances. The 1998–99 survey indicates that access to electricity increases the survival probability of newborns (up to one month old). The 1992–93 survey finds that access to electricity reduces under-five mortality risks, independent of the influence of clean cooking fuels.

Ridder and Tunali (1999) obtain similar results using Malaysian data after controlling for potential confounding factors, such as income. Using cross-country data constructed from comparable Demographic and Health Surveys, Wang (2003) finds a robust impact of access to electricity on under-five mortality, controlling for income and health expenditure.

These findings are difficult to interpret, because the survey instruments do not provide additional information that might link access to electricity and health outcomes. One possible explanation might be that a household connection to electricity facilitates access to information through television and radio, which are critical sources of information on public health; electricity also provides light for reading in the evening. Among high-income households, the health benefits of access to electricity may include refrigeration, which has been identified as an important measure for reducing the incidence of food-linked infectious diseases among young children.

Age-related risks. The China and India studies show that the impact of environmental conditions on child survival probability varies by age. The China study shows that access to safe drinking water sources significantly reduces child mortality risks after (but not before) the age of one month. Findings based on the India 1998–99 NFHS show that access to sanitation facilities (a flush toilet, pit toilet, or latrine) reduces child mortality risks for children between one and five but not those under one. Using survey data from Uttar Pradesh, Bhargrava (2003) shows that access to sanitation facilities significantly reduces infant mortality.

Studies from both sets of NFHS data from India (1992–93 and 1998–99) provide evidence that the use of clean cooking fuels (biogas, electricity, liquefied petroleum gas, kerosene, and charcoal) reduces child mortality risk. Using a hazard model that allows an age-varying effect of environmental conditions on mortality risks, Van der Klaauw and Wang (2004) find that having a separate kitchen and using clean cooking fuel significantly improves a child's probability of survival during the first month of birth but not thereafter. This finding is consistent with results from studies on health problems related to indoor air pollution in many low-income countries, which establish linkages between pregnant women's exposure to indoor air pollution and low birthweight and associated perinatal conditions of their children (Boy, Bruce, and Delgado 2002). (Health risks related to indoor air pollution are discussed in more detail later in this section.)

Child Morbidity from Diarrheal Disease

A large body of literature attempts to estimate the impact of access to safe water and sanitation on diarrheal morbidity. (Fewtrell and Colford 2004 provide a useful summary of the literature.) In estimating the global burden of disease from unsafe water and lack of sanitation, the WHO (Pruss-Ustan and others 2004) relies both on studies based on household survey data (for example, Esrey's 1996 analysis of data from the Demographic and Health Surveys) and randomized trials of home drinking water disinfection and hand washing.

Randomized trials of home drinking water disinfection have been shown to reduce the incidence of diarrhea in children under five by 44 percent in Bolivia and 62 percent in Uzbekistan (Semenza and others 1998; Quick and others 1999, 2002). Jalan and Somanathan (2004) show the importance of providing information about drinking-water contamination in inducing households to purify their water (box 3.1).

BOX 3.1
Does Health Information Increase Households' Efforts to Purify Water?

Jalan and Somanathan (2004) analyze the effect of information on household water-purifying behavior using a random-experiment approach. They find that households in Gurgaon, India, that were told that their drinking water was "dirty" were 11 percentage points more likely to begin some form of home purification in the next seven weeks than households that received no such information.

A water test kit, which costs less than $0.50 per sample, is available off the shelf from many nongovernmental organizations in Delhi and simple enough for households to use themselves.

The study shows that the impact of a water test kit on the probability of purification is about 25 times that of an additional year of schooling and more than two-thirds that of a move from one wealth quartile to the next. This result suggests that public programs that focus on disseminating health information are cost-effective and relatively easy to implement in low-income countries. Such efforts can stimulate demand for better environmental quality through political expression or increased willingness to pay for improvement of environmental services.

Hand washing is another area in which randomized trials have demonstrated the effectiveness of a simple method of reducing exposure to environmental contaminants. In a study conducted in Karachi, Pakistan, Luby and others (2005) find that children under five in households given plain soap had a 50 percent lower incidence of acute respiratory illness and a 53 percent lower incidence of diarrheal disease than children in control households. Cairncross and Valdemanis (2006) report similar results.

Using 1998–99 NFHS data from India, Jalan and Ravallion (2003) find significantly lower prevalence and shorter duration of diarrhea among children under five living in households with piped water. These gains are smaller for children with less educated women in the household, suggesting that education may be a proxy for knowledge about ensuring that water is safe to drink and that diarrheal disease is identified and treated in a timely manner. Access to an inside tap has a significantly larger effect on the duration of diarrhea in households in which the female member is illiterate, suggesting that an inside tap may partly compensate for the knowledge disadvantages of being illiterate.

Child Malnutrition

The nutritional status of young children is a function of household-level decisions regarding food consumption (quality and quantity), health outcomes, and child care. These choices are, in turn, determined by households' preferences, their access to health and basic environmental services, and their ability to utilize private as well as community resources (Alderman, Henschel, and Sabates 2003).

The immediate causes of malnutrition—including insufficient intake of energy, nutrients, or both—and the prevalence of infectious diseases are well known (Pinstrup-Andersen, Pelletier, and Alderman 1997). A large body of literature also relates malnutrition to diarrheal disease (Brown 2003). Children who experience repeated episodes of diarrhea are likely to become malnourished; water and sanitation interventions that prevent diarrhea may thus also prevent malnutrition.[3]

Previous studies by social scientists have concentrated on the impact of several underlying determinants—in particular maternal education, access to health care, and basic environmental services—on children's nutritional status (Barrera 1990; Thomas, Lavy, and Strauss 1996; Alderman and Garcia 1994). Glewwe (1999) shows that maternal education influences nutritional outcomes through several channels, including by directly transmitting health knowledge to mothers; teaching quantitative and literacy skills needed for the diagnosis and appropriate treatment of common childhood illness; and exposing women to modern medical treatment. Improving female access to education also improves the welfare of their children, by increasing women's control over resource-allocation decisions within households and increasing the use of health care services (Smith and Haddad 1999).[4]

Several studies provide evidence that private investment in female education and access to water and sanitation services is likely to generate external health effects on child nutritional status. These studies show that children living in households with inadequate access to basic services can still benefit from a neighbor's investment that results in better community environmental conditions. Studies using household surveys find similar evidence in several countries, including Brazil (Thomas and Strauss 1992); Ethiopia (Silva 2005a); rural Guatemala (Gragnolati 1999); Morocco (Glewwe 1999); and Peru (Alderman, Hentschel, and Sabates 2003).

Using the 1997 Peru Living Standards Measurement Survey (LSMS), Alderman, Hentschel, and Sabates (2003) estimate the derived health demand function, in order to study the external benefits of investment in access to education, water, and sanitation on children's nutritional status (measured by height for age). They find significant externalities to the investment in household-level environmental infrastructure (water and sanitation) and human capital (particularly female education in rural areas). In addition, they find that in rural areas, households with neither water nor sanitation infrastructure benefit only from being located near households with access to both services, not to safe water or sanitation alone. These findings are similar to those of Hughes, Lvovsky, and Dunleavy (2001) on child mortality in India.

The study by Silva (2005a), which uses model specification and estimation methods similar to those used by Alderman, Hentschel, and Sabates (2003), focuses on the impact of externalities of water and sanitation services on nutritional status, using the 2000 Ethiopia Demographic and Health Survey. She examines two nutritional indicators: underweight (weight for age), which is often regarded as a short-term measure of nutritional status, and stunting (height by age), a long-term measure. Two interesting findings emerge from the study. First, access to water and sanitation services has a significant effect on short-term nutritional status (underweight) but not on stunting. This result differs from the findings of Alderman, Hentschel, and Sabates (2003). Second, households' own access to water or sanitation has no significant impact on child health status (underweight); a strong health benefit emerges solely from community access to water or basic sanitation facilities.

While the external benefits of water and sanitation service coverage are important, these benefits are subject to declining returns (box 3.2).

These results on the external benefits of access to water and sanitation lend support to the "total sanitation" concept the World Bank and its partners have been applying in South Asia (World Bank 2005). This approach combines increased access to water and sanitation with public education on hygiene, as well as promotion of toilet usage through community action programs.

BOX 3.2
Diminishing External Benefits from Community Coverage of Water and Sanitation

The external impact of community water and sanitation conditions on children's health status depends on the average level of community access. In rural Peru the positive externality of access to sanitation services on health diminishes as the average level of community access to sanitation increases; the positive externalities on children's height become insignificant after about half of the neighborhood has access to sanitation (Alderman, Hentschel, and Sabates 2003). Silvia (2005a) finds diminishing external effects of community access to water and sanitation on underweight in Ethiopia. Hughes, Lvovsky, and Dunleavy (2002) show that the child survival probability is a function of the community level of water and sanitation access. They estimate that the critical threshold is about 50–60 percent of households with access to a private safe water connection or toilets, above which no additional health benefit is observed from infrastructure investment. The diminishing health impact of community-level environmental infrastructure is consistent with the finding by Jacoby and Wang (2004) that declines in child mortality in China increase when public investment targets safe water in poor localities, where environmental conditions tend to be much worse than in less poor areas.

Health Risks Induced by Indoor Air Pollution

Indoor air pollution poses a major health risk, in particular for poor households. Biomass fuel—such as wood, charcoal, crop residuals, and dung—remains the principal source of energy for cooking and heating in many rural areas of low-income countries. Many studies have confirmed a consistent statistical association between exposure to indoor air pollution and the incidence of diseases (acute respiratory infections, middle-ear infection, chronic obstructive pulmonary disease, lung cancer, and asthma) and a variety of perinatal conditions, possibly as a result of maternal exposure during pregnancy (Smith and others 2000; Ezzati and Kammen 2002; Smith, Mehta, and Maeusezahl 2004).[5] In low-income countries, acute respiratory illness caused by exposure to indoor air pollution is a leading cause of death among children under five; exposure to air pollution (both indoors and outside) during childhood can also have adverse health consequences into adulthood (Gauderman 2004).

The magnitude of the disease burden associated with the health risks of indoor air pollution is widely documented in the literature and increasingly recognized as a major health issue among health and environment experts. The WHO estimates that 1.62 million annual deaths and 38.54 million disability-adjusted life years (3 percent of the total) are associated with exposure to indoor air pollution (Smith, Mehta, and Maeusezahl 2004).

There are important differences between studies of the impact of water and sanitation on child health and studies of the effects of indoor air pollution. Most water and sanitation studies directly measure the impact of an intervention—hand washing, home drinking- water disinfection, or piped water connections—on health. In the case of indoor air pollution, there are fewer studies of interventions. It is possible to measure indoor concentrations of particulate matter (PM10 or PM2.5) and even to measure personal exposure to particulate matter, which can be related to mortality and morbidity through a dose-response function. It is also possible to study separately the factors that affect indoor air pollution concentrations.[6] There is a growing body of literature on the impact of fuel use, stove type, and other factors on indoor air concentrations (Ahmed and others 2005; Dasgupta and others 2004, 2006a). There is also a growing epidemiological literature on measurements of personal exposure to health risks.[7]

Ezzati, Saleh, and Kammen (2000) and Ezzati and Kammen (2001) estimate dose-response functions for indoor air pollution. In a series of studies in Kenya, they measured concentrations of PM10 inside 55 homes and recorded the time spent by different family members indoors, in different parts of the house, and outdoors, using these data to construct measures of personal exposure. They then performed a cross-sectional analysis relating the incidence of respiratory illness, diagnosed by health professionals over a two-year period, to personal exposure. Ezzati and Kammen find that a child exposed to 24-hour concentrations of particulate matter of 1,000–2,000 micrograms per cubic meter is more than twice as likely to experience acute lower respiratory infections than a child exposed to 24-hour concentrations below 200 micrograms per cubic meter.[8]

Three other important findings emerge from this work. First, for the highest-exposure group (the women in charge of cooking and the young children they look after), about half of daily exposure occurs in a high-intensity episode (cooking period). Second, the oft-reported significant gender effect disappears when controlling for time spent for cooking and period of high-intensity indoor air pollution exposure, suggesting that the gender variable simply picks up the effect of omitted cooking time and peak exposure variables. Third, empirical results based on average daily PM10 concentration measures can significantly underestimate the relative exposure of women, resulting in a systematic bias in the assessment of the exposure-response relationship.

To better understand the factors determining personal exposures, Dasgupta and others (2006b) measured indoor air concentrations using newly developed monitoring equipment. They used air samplers that measure 24-hour average PM10 concentrations and real-time monitors that recorded PM10 and PM2.5 concentrations at 2-minute intervals for 24 hours for a stratified sample in urban, periurban, and rural areas of the Dhaka region of Bangladesh. The exposure measures focus on two dimensions: an individual's time spent in different locations (cooking areas, living areas, and outside) and hourly fluctuations in pollution from cooking.

Three main findings emerge from this study. First, indoor air pollution is not confined mainly to cooking areas but can rapidly disperse into living areas (where monitoring data show that pollution is only moderately lower than in cooking areas). Second, infants and children under five of both genders face a high level of exposure (about 200 micrograms per cubic meter). The gender-based divergence occurs among adults, with women's exposure nearly twice that of men in the 20–60 age group and about 40 percent higher for women over 60. Third, poorly educated women in poor households face indoor air pollution levels that are four times those of men in higher-income households with more educated women. These findings are consistent with those of a study of indoor air pollution exposure in Andhra Pradesh in India (World Bank 2002).

The Bangladesh studies suggest that health risks related to indoor air pollution can be reduced by improving ventilation and reallocating activities and time spent indoors during high-emission periods.[9] For example, children's exposure to indoor air pollution can be halved by simply increasing their outdoor time from three to five hours a day and concentrating outdoor time during peak cooking periods. This finding suggests that the primary focus of policy should be providing information to households to influence their allocation of activities; encouraging investment in cleaner stoves or use of cleaner fuels should be of secondary concern, particularly in the short term.[10]

How Robust Are the Empirical Findings?

The inherent difficulties in studying the determinants of health outcomes based on household survey data raise questions about the robustness of the empirical findings. This is a highly relevant point, as empirical results from these studies are often used as the basis for policy making, ranging from allocating public investment across sectors (for example, among health, education, the environment, and energy) to making investment choices among different types of environmental infrastructure to targeting various health-focused public programs.

Empirical findings from cross-sectional household surveys are often criticized on the ground that they fail to prove a causal relationship between health and environmental conditions and provide biased estimates of the impact of environmental variables (for example, access to safe water sources or use of cooking fuels).[11] Critics argue that these studies do not provide useful guidance for allocating resources for public infrastructure investment or health programs. This deficiency in household surveys has led to a tendency in the recent economic literature to endorse only findings from randomized trials or properly designed experimental field studies.

Despite the obvious advantages of randomized trials or experimental approaches, studies on the determinants of health outcomes are likely to continue to rely primarily on household surveys, for several reasons. First, because of the rarity of child mortality, measuring its risk requires a large sample or the accumulation of mortality experiences of smaller samples over long periods (often more than five years). Second, health outcomes of exposure to environmental risks may depend on cumulative exposure; it is therefore very difficult to apply short- or medium-term program evaluation approaches to assess the health impacts of a program. Third, although is possible to implement randomized trials for some environmental interventions (such as improved cooking fuels, improved stoves, water projects, or nutritional programs), the difficulties of randomizing infrastructure services (piped water and sanitation services) make it difficult to apply experimental approaches.

Despite the analytical constraints of various approaches for studying the linkages between health and environmental conditions, the studies reviewed in this chapter share several robust characteristics. First, the China and India studies on child mortality draw on large household data sets that provide a sufficiently large number of child deaths for analyzing child mortality risks.

Second, the China study validates the causal effect of environmental factors on child mortality using cause of death information. The results from the hazard function, which allows the child mortality rate to vary by cause of death, confirm that access to safe water does not affect the probability of death from causes such as birth-related deaths or neonatal tetanus, which should not be associated with safe water. These findings are very similar to those Galiani, Gertler, and Schargrodsky (2005) obtain using municipality mortality data from Argentina. They find that increases in access to the water and sanitation network and improvements in service quality following privatization were associated with a significantly lower rate of child mortality from infectious and parasitic diseases and perinatal deaths, but no significant change in mortality from other causes, such as accidents, cardio-vascular disease, or cancer.

Third, by allowing environmental effects on child mortality risk to vary by age, the China and India studies show that environmental factors (safe water access or sanitation facilities) that are not likely to affect neonatal deaths are indeed not significant. The fact that spurious correlations between health and environmental conditions are not picked up increases confidence in these studies.

Confidence in the studies on nutritional status is bolstered by the fact that results for several countries are very consistent (in particular those of Alderman, Hentschel, and Sabates 2003 on Peru and Silva 2005a on Ethiopia). The critical issue in nutritional studies is the sample selection bias resulting from using only the survival population, as discussed above. This remains an important area for future research, particularly in African countries with high mortality rates.

Relatively few studies have been conducted on the health effects of indoor air pollution. Policy conclusions regarding health impacts of different levels of exposure and methods to reduce exposure must await the availability of good-quality monitoring data and the application of appropriate analytical methods in a larger number of studies.

Conclusions and Tentative Policy Implications

Empirical problems in analyzing the linkages between health and the environment and the sometimes weak evidence from the literature suggest that it is important to interpret results from studies based on cross-sectional household survey data cautiously. The findings nevertheless provide useful policy implications for guiding project or program design in four areas.

First, the research reinforces the message that designing health-focused programs and projects should be based on much broader considerations, including health, the environment, education, nutrition, and public health information. How to design and implement such multidimensional programs remains a challenge. Future policy and analytical work should aim to provide more specific and operational guidance for policy lending and project investment that aims to address health issues.

Second, allocating resources, in the form of public programs or direct public investment in environmental infrastructure, should focus on targeting poor communities rather than poor households, because investments in clean water and sanitation infrastructure have an externality effect on household health. Publicly funded programs need to recognize and capture this externality.

Third, the role of information, which plays a critical role in improving health outcomes in low-income countries, has been largely overlooked in many health-related studies. More important, the role of health information is often ignored in policy making, which may imply that health program resources are being misallocated. The lack of information about the health impacts of poor environmental services—ranging from water quality to exposure to indoor air pollution—may affect the demand for better environmental quality (by way of political expression or lack of willingness to pay for improvements) and household behavioral responses in mitigation. Future studies on environmental health should focus on evaluating the impact of public information on household mitigation behavior and health outcomes. The role of information has important implications for guiding health-focused program and project design.

Fourth, while it is widely recognized that the use of biomass fuel poses a serious health risk to households in low-income countries, the factors that determine human exposure and policy recommendations on reducing exposure require further study. Key factors include energy technology (high-efficiency and low-emission stoves); housing characteristics; and behavioral responses (who is assigned to cooking tasks within households and the amount of time spent indoors during peak cooking periods). More thorough cost-benefit analysis is needed to provide answers to such questions as whether public programs should focus on promoting wood stoves or the transition of fuel use from biomass to charcoal, kerosene, or gas. In addition, studies on the health effects of indoor air pollution should consider linkages between fuel use, deforestation, and carbon emissions, in particular when climate change has become a policy focus.

Future policy analysis should focus mainly on improving data collection to address deficiencies and enhance the robustness of empirical evidence. In particular, researchers should attempt to collect longitudinal survey data and to incorporate questions in household surveys on cause of death and other retrospective information on social, environmental, and health conditions at the household level. Evidence generated from household surveys should be validated by studies using experimental approaches in appropriate circumstances or matching methods to control for unobservable confounding factors.

Notes

1 In the epidemiological literature, case-control studies are often conducted to study rare outcomes. For example, to study the impact of indoor air pollution on deaths from acute respiratory illness, a sample of children who die from exposure to indoor air pollution would be compared with a control group of children who die from other causes (such as diarrheal disease). Case-controlled studies, however, suffer from omitted variable bias and rarely include enough observations to apply techniques such as propensity-score matching.

2 Hughes, Lvovsky, and Dunleavy (2001) and Van der Klaauw and Wang (2005) include piped water connection, hand pumps, and wells located in a household's yard or inside dwelling as private sources of safe water.

3　The WHO estimates that 50 percent of the burden of disease associated with malnutrition is attributable to environmental factors (WHO 2002).

4　Women tend to allocate a larger share of the household budget to children's food consumption than do men, who allocate a larger share to alcohol or cigarettes.

5　The current scientific consensus indicates that most respiratory health problems result from inhalation of respirable particles with a diameter of less than 10 microns (PM10) and particularly finer particles (PM2.5) released from combustion of solid fuel.

6　It is also possible to study the impact of interventions such as the introduction of improved stoves using experimental methods (see Smith 2006).

7　In the early literature relating indoor air pollution to child health (Smith and others 2000), exposure is usually measured very crudely—with dummy variables indicating type of fuel used for cooking, for example.

8　This result controls for child age and gender but not for nutritional status, which was not measured in the study.

9　Pitt, Rosensweig, and Hassan (2006) use a panel data set constructed from the 1981/82 and 2002/03 Bangladesh household surveys to test the assumption that an individual's health endowment affects his or her allocation of cooking time. Their study suggests that households rationally allocate cooking activities to women who are in poorer health. If this is the case, it may be difficult to reduce these women's exposures without improving ventilation or changing the type or amount of fuel burned.

10　Improved stove programs have not been a great success, partly because of lack of community involvement in stove design and possibly because of failure to understand the long-term health consequences of exposure to indoor air pollution. For example, the clean stove program of Enterprise Works in Ghana mentions nothing about the health effects of using improved stoves, emphasizing instead fuel savings, reduced deforestation, and reduced greenhouse gas emissions.

11　It is widely known, for example, that households with latrines behave more hygienically in general than households without latrines, making it difficult to attribute any health benefit to latrines alone.

CHAPTER 4

Household Welfare and Policy Reforms

POLICY CHANGES THAT AFFECT THE NATURAL ENVIRONMENT can have direct and indirect impacts on household welfare. Poverty alleviation and an increase in a household's economic welfare are one set of possible impacts. Better nutritional and health outcomes are another possible effect. This chapter focuses on policy reforms that affect both aspects of household welfare through better management of environmental resources.

Reforms with positive environmental and welfare impacts do not always originate in the environmental sector. Some reforms—such as creation of common property rights, incentives for better management of natural resources, or novel markets for environmental services—pertain directly to environmental resources. In other cases, sectoral or macro policies intended to improve other aspects of the economy may also have environmental and welfare benefits (strengthening of private property rights is one example).

The past several years have seen significant changes globally in who has access to and control over natural resources. There have been parallel trends toward strengthening the rights of local communities and the private sector over natural resources in many countries. The strengthening of local rights has been achieved through devolution of state control to communities, increased legal access to natural resources, and power-sharing agreements with the state. The strengthening of private rights has occurred thorough privatization of public sector enterprises, improved security of land tenure, and the creation of economic value from environmental services.

Many reforms fail to accomplish their goals or have unintended consequences for the poor. In some cases, for example, strengthening communitarian rights may exacerbate the deprivation of the very poor. Sarin and others (1988), Sundar (2000), and Agarwal (2001), for example, argue that by closing off access to certain forests, joint forest management programs in India help well-off villagers, who can secure alternate sources of fuelwood, but burden poor villagers and women, who cannot. Dzingirai (2003) argues that community-based natural resource management programs such as Communal Areas Management Program for Indigenous Resources (CAMPFIRE) in Zimbabwe do not benefit the rural poor.

Similarly, extending private rights and creating novel markets may not always benefit all of the poor. Securing land rights for one group may deny them to another. Munyao and Barrett (2006) find that more secure land rights had a negative impact on traditional pastoralism in northern Kenya. Brasselle, Gaspart, and Platteau (2002) show that less secure land tenure in Burkina Faso encouraged more investment in land where such investments could improve future tenure security. Zbinden and Lee (2005) find that when payments for environmental services are targeted to owners of large forest areas, program payments tend to go to better educated, wealthier farmers.

This chapter examines a range of policy reforms by presenting case studies that document positive benefits to the poor. The six cases assess the impacts of various policy reforms on broad indicators of welfare. The policy mechanisms highlighted in the studies were crafted to influence environmental resources—such as forestry, wildlife, water, and land—in Nepal, Namibia, the Philippines, China, and Nicaragua. The Argentine case study examines reforms aimed at increasing coverage of water supply and sanitation.

The rest of the chapter is organized as follows. The next section focuses on the policy reforms examined in the case studies. The third section draws attention to the need for the right kind of data collection and identifies some of the limitations of the case studies. The last section examines advances in knowledge of environment-poverty linkages derived from household-level welfare analysis.

Selected Policy Reforms: Evidence from Case Studies

This section focuses on five policy reforms: creating common property rights, strengthening private property rights, creating incentives for better management of environmental resources, creating novel markets for environmental services, and increasing access to services. Of these reforms, creating common property rights and creating incentives for better resource management are devolution-type policy reforms. Strengthening private property rights, creating novel markets for environmental services, and increasing access to services are policy reforms that build on private property rights.

Interest in privatizing water utilities stems from the desire to increase access to water infrastructure. In general, privatization of public sector functions has two goals: increasing efficiency and increasing access to new financial resources through private investment. This chapter looks at privatization as a reform tool that allows increased access to water infrastructure.

The case studies examined use impact evaluation methods to measure the impact of policy reforms on household welfare. Various methodologies are used (box 4.1).

Creating Common Property Rights in Namibia and Nepal

Creation of common property rights implies the transfer of rights and responsibilities from the state to user groups at the local level. In Namibia, for example, registration of communal conservancies provided communities with the rights and responsibilities of wildlife management within the conservancies. In Nepal changes in national forest policy allowed local user groups to manage forests.

Common property rights are by no means uniform across countries or types of properties. In Namibia formation of conservancies allows communities as a whole to enjoy use rights to wildlife, though these rights do not carry over to individual households. In Nepal individual households enjoy collection and use rights to fuelwood from community forests. The case studies of Namibia and Nepal highlight similarities and differences in policy reforms and the ways in which these reforms affect household welfare.

The first study (Bandyoadhyay and others 2004) focuses on increased legal access to wildlife through community conservancies in Namibia, where communities have certain rights over wildlife and tourism. The second study (Bandyopadhyay, Shyamsundar, and Kanel 2006) focuses on the devolution of control of forestry resources to community forest user groups (FUGs) in Nepal. Both studies consider the impact of community participation in wildlife and forestry management on consumption expenditure and other measures of household welfare.

Creation of common property rights does not always imply that the disadvantaged in the community have equal access to those properties. In particular, the issue of elite capture cannot be ignored in any discussion of common property rights. Elite capture is the situation in which a few elites in the community usurp the rights to a common property and exclude the disadvantaged from exercising their common property rights.

Some of the case studies test the elite capture hypothesis and more generally try to answer three questions:

- Who participates in the community management of environmental resources?
- Do participants gain more than, the same as, or less than the rest of the community?
- Are poor and disadvantaged households prevented from benefiting from the common properties?

BOX 4.1
Impact Evaluation Methods

Impact evaluation methods attempt to determine the average gain in welfare to households included in a reform compared with the hypothetical situation in which the same households were not included. The resulting measure is known as the impact of average treatment on the treated (ATT) (Imbens 2004).

The hypothetical welfare of households that were not included in the reform cannot be observed. Empirical analyses depend on dividing the sample households and communities into control and treatment groups. Randomized social experiments would ensure that households participating in the reform are not statistically different from those in the control group. However, randomized social experiments imply denying the benefits of reform to some households that may need it most. In some cases—including the cases examined in this chapter—the nature of the reform may make randomization at the household level impractical (Moffitt 2003; Keane 2006).

In cases where the nonrandom allocation of the treatment is either determined by the policy maker or self-selected by households, selection bias may cloud the impact estimation results. Selection bias may be of two types: bias based on observed data and bias based on unobservable data. The difference in method takes into account selection biases of both kinds by taking the difference between the average welfare of the treatment and control groups before and after implementation of the reform. This method requires data from before and after program implementation for both the control and treatment groups. This method was used in the Argentina case study.

In the absence of before and after data, estimation methods are limited to cross-sectional analytical tools. Most of the studies examined in this

---

Box 4.2 focuses on the first two questions, drawing on similarities and differences across studies. Box 4.3 looks at who—the poor or the nonpoor—benefits more from the reforms.

Namibia. Namibia has pioneered legal access of communities to wildlife resources through communal conservancies. Its community conservancy program was largely shaped by the presence of commercial conservancies that formed a successful wildlife industry (Jones and Murphree 2001). In 1995 the government of Namibia laid out a set of progressive access rules for communal lands.[1] Under the policy reform, communal conservancies as a whole could exploit and gain from wildlife resource management. Few studies have quantitatively assessed the welfare impact of Namibia's communal conservancy program. Jones (1999b) provides anecdotal evidence that communities have benefited in cash and kind.

The policy reform in Namibia requires that communal conservancies register with the state, providing it with recorded geographical boundaries and a comprehensive list of members. Communities in registered communal conservancies enjoy chapter used cross-sectional data. The propensity score–matched difference method calculates ATT differences in welfare between treatment and control households after they are matched with one another on the basis of propensity scores (Imbens 2004). Each household is assigned a propensity score based on a vector of its characteristics. The propensity score–matched difference method can correct for selection bias based on observed variables if these variables are included in the calculation of the propensity score. This method is not appropriate if selection biases based on unobserved variables are present.

The cross-sectional data in the case studies were used with the instrument variable method to calculate ATT (Wooldridge 2002). In this method, both selection into the treatment group and the welfare indicator are modeled with estimated parameters of equations. Unlike the estimates derived from other methods, the estimates yielded from this method depend on the structure of the models. Like the difference in difference method, the instrument variable method does not suffer from the two types of selection biases based on observed and unobservable factors.

The inherent problems of using cross-sectional data mentioned in the previous chapters are present in the case studies examined here. In particular, cross-sectional data can identify only associations between a policy change and its possible effect on an outcome. Without the time dimension in the data, analytical methods alone cannot determine causalities where the factors are confounding. The associations identified in the studies may reinforce or point in the direction of possible relationships, however.

For a comprehensive review of household welfare–based impact evaluation methodologies, see Ravallion (2007).

economic rights to wildlife resources within the boundaries of the conservancies. The communities also take responsibility for conserving these resources. By the end of 2003, 23 percent of all communal land in Namibia was under conservancies (NACSO 2004).

The communal conservancies prepare annual wildlife management plans that include a count of existing stock. Their allocated use is subject to state regulations for protection of understocked species. Jones indicates that meat distribution to member households is a major benefit. Communal conservancies may gain from profit-sharing agreements with tourist lodges and employment generated through tourism-related activities.

The case study on Namibia (Bandyoadhyay and others 2004) used household survey data collected in 2002 by the Wildlife Integration for Livelihood Diversification (WILD) project and the Environmental Economics Unit of the Directorate of Environmental Affairs in the Ministry of the Environment and Tourism. It included 1,192 households in seven conservancies from two regions, Kunene and Caprivi.

BOX 4.2
Who Participates in the Community Management of Environmental Resources?

In some cases and countries, when resource management is handed over to communities, all households automatically become members of the local institution. However, even under these circumstances, not all households engage with the community organization or even know about it. How important is it for households to actively participate? Does lack of participation reduce benefits, given that most changes affect the entire community?

In Namibia community conservancies increased the welfare of households living within them, but participants did not gain relative to nonparticipants. Bandyoadhyay and others (2004) speculate that participants may share their wildlife-related income with others. It is also possible that the increase in welfare was a result of community public goods and NGO activities in the area.

In Nepal households' participation in forest user groups was not observed in the data. However, a similar study on joint forest management in India (Bandyopadhyay and Shyamsundar 2004) finds that community forestry did not increase access to fuelwood consumption for the community as a whole, but it did increase fuelwood consumption by households that participated in the program. It is possible that participation translated into greater access to forestry officials and increased local power. The results from the India study, which may not be applicable in Nepal, are the opposite of the results of the Namibia study.

Ideally, impact estimations should be based on randomized experiments. Randomized experiments may not always be practical in the case of many environmental management policy changes, however. The level and intensity of household participation as a result of the policy change toward participatory environmental management may differ across sectors and countries. It is important not only to understand what motivates households to participate in community- and private-entity-based environmental management but also to measure how participating households stand to gain from participation.

The survey did not include households living outside the seven conservancies. To overcome this data limitation (the lack of a proper control group), the study used the fact that the full benefits from a conservancy can be achieved only after the conservancy has been in operation for several years. It thus distinguished between two types of conservancies, established and recent (as comparator). It then evaluated differences in income measures between these two types of conservancies. This study used instrument-variable and propensity-score-based impact estimation methodologies. These methods econometrically compare households in control groups with those in treatment groups (see box 4.1).

BOX 4.3
Who Benefits Most from Community Management?

Elites in the community may take over the management of and reap the potential benefits from environmental resources, excluding the rest of the community. Alternatively, the gains from the resources may be evenly distributed within the community, and participation in the management initiatives may not result in additional benefits to participating households.

Does policy reform contribute to some form of elite capture? Are richer and poorer households equally well off as a result of increased community or private control? Does some community management hurt the poorest people in the community?

Identifying the differential distributional impacts on the poor versus the extremely poor or smallholders versus the landless is important to policy makers. By identifying vulnerable subgroups in the community that enjoy fewer benefits than others, studies can facilitate better targeting of programs.

The Nepal study distinguishes between land-nonpoor (households in the top three quintiles with respect to the value of land held); land-poor (households in the bottom two quintiles with respect to the value of land held); and landless households. It finds that the land-poor gain more from community forestry than the land-nonpoor or the landless. This finding suggests that the land-poor are better off when forests complement existing private assets. Landless households are at least as well off as the land-nonpoor in terms of welfare gains from community forestry.

The Nicaragua case study looks at nonpoor, moderately poor, and extremely poor participants to the PES program. The study shows that moderately poor households, but not the extremely poor, consistently participate more and reap greater benefits than the nonpoor and the extremely poor. Extremely poor households are at least as well off and participate at least as much as nonpoor households. An important issue that needs to be probed is thus how institutional changes affect the needs of the landless rural poor in Nepal and extremely poor PES participants in Nicaragua.

The case study on Argentina finds that privatization of water systems did not affect child mortality in municipalities in which less than 25 percent of households were poor. In contrast, in municipalities in which more than half of all households were poor, privatization was associated with a 26.5 percent reduction in child mortality

All six studies find that the poor, the vulnerable, and other disadvantaged groups (including the less educated households and households headed by women) benefited at least as much as the rest of the community from the devolution of management of natural resources to communities (in the cases of Nepal, Namibia, and the Philippines). The poor benefited more than the nonpoor from the devolution of environmental resource management to private entities in Argentina and Nicaragua.

Bandyoadhyay and others (2004) examine four indicators of welfare: household income, household consumption, per capita income, and per capita consumption. In the Kunene region of Namibia, treatment communities had higher per capita income (28 percent higher) than comparator groups, which was attributed to the presence of established conservancies. These improvements in income were attributed to an increased ability to engage directly with tourism as well as activities of nongovernmental organizations.

This study finds the impact of conservancies was poverty neutral in some regions and propoor in others. It examines the welfare impact of conservancies in four types of disadvantaged households: those with low education levels, those headed by women, asset-poor households, and livestock-poor households. In all cases, it finds that the disadvantaged groups were at least as well off as the rest of the communities in terms of benefiting from communal conservancies.

This study demonstrates that devolution of common property rights to community conservancies and increased economic activities resulted in measurable welfare gains to households. Moreover, the poor and other disadvantaged groups gained at least as much as other groups in the communities.

Nepal. Nepal is a prime example of institutional change in forestry. In 1993 the government passed a forest act that radically changed forest use (Kanel 2004). This act resulted in the transfer of nationalized forests from state control to local communities. FUGs were the institutional tool used to facilitate this transfer. Forest transfer to communities accelerated in the 1990s; as of 2000 some 13,000 FUGs managed 25 percent of Nepal's forests.

The government of Nepal has strengthened the rights of local communities over forests by signing power-sharing agreements, legalizing access to forests, and decentralizing forest oversight agencies (Shyamsundar, Araral, and Weeraratne 2005). The new policy requires that local communities create FUGs and register them with the district forest officer. The FUGs have responsibility for creating a forest operational plan for the community forest. Operational rules to protect, harvest, use, and manage the forest are under the control of the FUG. Local forest officers provide FUGs with technical advice on forest management, as well as seedlings for rehabilitation, and they help stem violations of rules and resolve conflicts among users (Tachibana and others 2001).

Unlike the restriction of communal resource use in Namibia, in Nepal FUGs can and do allow individual households access and use of the forestry resources for domestic fuelwood consumption. The study assesses whether community forestry and greater household access to common resources translates into household welfare gains. It hypothesizes that greater community property rights over forest assets and increased access to funds for infrastructure development and services should result in improvements in household consumption and household income.

The study by Bandyopadhyay, Shyamsundar, and Kanel (2006) uses data from the Nepal Living Standard Survey (NLSS II) conducted by the Central Bureau of Statistics between April 2003 and April 2004. It follows the World Bank's Living Standard Measurement Survey methodology. It examines rural households in three regions where forestry user groups are common: the rural western mountains and hills, the rural eastern mountains and hills, and the rural western Terai.

The treatment group consists of households and communities that had formed FUGs; the control group includes households and communities that did not participate in community-based forest management. Impacts of FUGs are estimated at both the community and household levels. At the community level, semiparametric propensity-score-based methods are used to measure the impact of FUGs on fuelwood collection. Parametric methods of maximum likelihood estimation of a two-equation model are used to estimate the impact of FUGs on household fuelwood collection, income, and expenditure measures of welfare.

Using data from three districts in Nepal, Edmonds (2002) robustly shows that community forestry resulted in a 14 percent decline in fuelwood collection. Bandyopadhyay, Shyamsundar and Kanel (2004) find no measurable difference in fuelwood collection between FUG and non–FUG villages. One possible reason for the differences may be that the data Edmonds used were from the early days of FUGs, when forests were depleted, while the data Bandyopadhyay, Shyamsundar and Kanel (2004) used represent more established FUGs, with regenerated forests.

Bandyopadhyay, Shyamsundar, and Kanel (2004) examine the nature of the impacts of community forestry. They examine whether community forestry, by increasing local control over forest resources, improves household welfare. Over time, community management of forests is expected to increase household income by increasing the biomass available from forests; increasing the stock of agricultural and livestock inputs obtained from forests; reducing labor time used for collection activities; or improving the flow of services provided by forests. The study finds that the presence of community forestry and reinvestment in community infrastructure in a village is associated with a 6 percent increase in household welfare.

Strengthening Private Property Rights in China

Jacoby, Li, and Rozelle (2003) suggest that insecure land tenure in China has prevented much needed investment in land improvement and may have contributed to environmentally unsustainable methods of cultivation and over-exploitation of natural resources.

Adoption of household land-use rights under the household responsibility system in the late 1970s and early 1980s in rural China contributed to agricultural productivity gains (Lin 1992). However, agricultural growth flattened out in the late 1980s. It is widely believed that more secure individual land-use rights could improve agricultural growth (Qi 1999). In 1986 the government revised the Land Management Law to improve tenure security, extending land tenure to 30 years.

The China case study examines the impact on long-term investment in land by private land users of an experiment in longer land tenure. In 1987 the State Council designated Guizhou Province as an experimental area. Agricultural land-use tenure in the province was extended to 50 years—20 years longer than the 30-year national tenure in 1994. Guizhou Province also stopped the practice of adjusting the size of landholdings based on population changes. These two measures provided a higher level of security of tenure to farmers in Guizhou Province than in the rest of China.

To explore the impact of longer land tenure, Deininger and Jin (2003) used survey data on 1,001 households from 110 villages in Guizhou, western Hunan, and Yunan Provinces. Hunan and Yunan Provinces were chosen as control areas on the basis of their proximity and climatic and geographical similarity to Guizhou. They find that longer land tenure in Guizhou Province is associated with investment that is 2.6–2.8 times greater than in the control provinces. Investments in long-term sustainable agricultural practices are more profitable when tenure rights are ensured. Such investments are also environmentally sustainable and may include positive environmental externalities to the community, such as better watershed management.

Creating Incentives for Better Management in the Philippines

Management of irrigation water resources has traditionally been the responsibility of the state. The earliest transfers of the management of irrigation water services to farmer organizations took place in the United States and France in the mid-20th century. Governments in Africa, Asia, and Latin America have reduced their roles in irrigation management, ceding them to irrigation associations and farmers groups (Vermillion 1992). According to Vermillion (1997), irrigation management transfers (IMTs) are preferable to centrally managed systems for three reasons: (a) farmers have direct interests in managing irrigation systems, while state bureaucracies may not have the right incentives; (b) an increase in efficiency from IMT may offset any increased cost of irrigation to farmers; and (c) IMT saves government resources by limiting their responsibilities for routine operation and maintenance.

Along with irrigation management responsibilities, farmers groups or irrigation associations may also be allowed to collect irrigation fees and retain part of the fees to offset operations and maintenance expenditures. While early IMTs were targeted to large-scale farmers in developed countries, recent IMTs in developing countries have targeted poor and small-scale farmers. IMTs as donor-funded projects have gained ground in recent years (Groenfeldt and Svendsen 2000; Shah and others 2002).

The case study of the Philippines (Bandyopadhyay, Shyamsundar, and Xie 2007) focuses on power-sharing agreements between the state and user groups in the form of IMT contracts. It assesses impacts on maintenance efficiency and farm yield. Fifty percent of the irrigated area in the Philippines is managed publicly under national irrigation systems, 37 percent is managed by communal irrigation systems, and 13 percent is managed by private irrigation systems. The national systems are owned and operated by the National Irrigation Administration, a semiautonomous government corporation that is responsible for irrigation development (Bagadion 2002; Sabio and Mendoza 2002). In the late 1990s, the National Irrigation Administration initiated IMT contracts with selected irrigators associations that handed over irrigation fee collection and operations and maintenance responsibilities of secondary canals to irrigators associations.

The study area included 1,020 farm households in the Magat River Integrated Irrigation System (MRIIS) in Region 2 in Luzon. The irrigation system is located in the basin of the Magat River, which runs into the Cagayan Valley. The study compares the performance of a random sample of 43 treatment irrigators associations under IMT contracts and 25 control irrigators associations not under IMT. The focus is on rice production in areas in which power-sharing agreements in the form of IMTs had occurred between farmer organizations and the national irrigation agency. The study finds that IMT gave irrigators associations greater access to resources through member fees and that allowed them to more directly respond to maintenance requirements and to control the release of water. In IMT areas where farmers effectively managed resources, rice yields were 2–6 percent higher than in non–IMT areas, even after controlling for various other differences.

Participating in Novel Markets in Nicaragua

Payments for environmental services (PES) have emerged as a novel market mechanism to finance conservation in developing countries (Landell-Mills and Porras 2002; Pagiola, Landell-Mills, and Bishop 2002; Wunder 2005). PES is based on two principles: those who benefit from environmental services should pay for them and those who contribute to generating these services should be paid.

The PES approach has three potential advantages: (a) it accesses financing sources that may not otherwise be available for environmental management; (b) it may be sustainable, if its incentives are compatible for both service users and providers; and (c) it may be efficient, in the sense that it would only work for environmental services whose benefits exceed the costs to service providers. However, for global environmental services, such as conserving biodiversity and sequestering carbon dioxide, PES may depend on donor funding and may compete with other donor-funded activities.

Pagiola, Arcenas, and Platais (2005) raise three key questions regarding potential linkages between PES and poverty: Who participates in PES, and how many of them are poor? Are poor households able to participate in PES programs? Are poor households affected indirectly by PES programs?

The Nicaragua case study (Pagiola, Rios, and Arcenas forthcoming) differs from the other case studies, in that it focuses only on participation in the PES program. In this case, the benefits to households in the form of PES are well defined and nonrandom. The study uses a variety of measures of participation based on the area under silvopastoral land management and the complexity and intensity of the program chosen by households. The area and chosen intensity of the program determined the amount of benefit payments to households. Thus the observed level of participation had a direct and proportional effect on received benefits. Households that did not or could not participate in the PES program enjoyed no benefits.

The study considers examines the Regional Integrated Silvopastoral Ecosystem Management Project implemented in Nicaragua and other countries as a pilot PES program financed by the Global Environment Facility. The silvopastoral practice includes three components: (a) planting high densities of trees and shrubs in pastures to provide shade and diet supplements and prevent soil erosion; (b) creating fodder banks in areas used for other agricultural practices; and (c) using fast-growing trees and shrubs for fencing and wind screens.

The study focuses on the participation of poorer households in the PES program, which promoted silvopastoral land use at various levels of technical complexity to livestock farmers. Compared with traditional pastoral land use, silvopastoral land use has several public externalities, such as better watershed management, increased biodiversity, and higher carbon sequestration. The PES program internalized some of these benefits and offered monetary incentives to private landholders for observable changes in land use owing to silvopastoral practices.

The study used before and after data from 2002 and 2004 for 103 households in the Matiguas-Rio Blanco area, located in the department of Matagalpa, about 140 kilometers from Managua. The analysis was conducted for three groups, the nonpoor, the moderately poor, and the extremely poor.

The authors finds that moderately poor households participated in the program to a greater extent than nonpoor households and thus benefited more from the program. The extent of participation by poor households was not limited to simpler and less expensive options. Moreover, poor households tended to implement more substantial changes in land use. By undertaking complex land-use changes, poor households provided greater environmental benefits and in return received higher payments. The study also finds that the intensity of participation for extremely poor households was not significantly higher than for nonpoor households (see box 4.3).

Increasing Access to Services in Argentina

A key target of the Millennium Development Goals agreed upon by UN member countries in 2000 was to reduce the number of households with no access to safe water by half by 2015. There is little consensus on how to increase access to safe water to a large part of the population. Privatization of water services is one potential method of doing so, although it is not without controversy. What follows is not about privatization per se. It looks at a particular privatization in Argentina to examine whether increasing access to safe water had a measurable impact on health outcomes.

Water services were managed by the federal water and sanitation authorities from the late 19th century until 1980. By 1990 local public companies provided water services to two-thirds of municipalities, and nonprofit cooperatives provided services to the remaining one-third. Privatization of public water services started in Argentina in 1991. By 1999 about half of all local public water companies had been turned over to private enterprises.

Galiani, Gertler, and Schargrodsky (2005) look at annual municipality-level child mortality and other data to estimate the impact of increased access to safe water as a result of privatization of water services in Argentina. The availability of annual time-series data for 1990–99 allowed the authors to analyze before and after data for both the treatment and the control groups of various municipalities. The study uses difference-in-difference estimations to measure the impact of privatization on child mortality (see box 4.1).

Local governments in Argentina that privatized water services were motivated by potential efficiency gains and savings in public expenditure; it was not clear that the increased efficiency gains from privatization would result in improved health outcomes. The study finds that privatization of water services was associated with a 5 percent reduction in the child mortality rate from the baseline.

This study is one of the few studies that finds significant health benefits from privatized water supply. In a meta-study of water privatization and public health in Latin America, Mulreany and others (2006) find no compelling case for privatizing public water utilities on public health grounds.

Challenges and Data Limitations

Household-level data are necessary to establish linkages between various environmental and natural resource management activities and household welfare. Collection of large-scale household data is expensive and time consuming. The six case studies in this chapter used household survey data from a variety of sources. Advantages and disadvantages are associated with each type of survey. This section examines some of the challenges and limitations of the types of data used.[2]

The Nepal study of FUGs and measurement of household welfare attributable to the community management of forestry resources used data from the Second Nepal Living Standards Survey, conducted in 2003–04. The case study was based on data from 1,708 households in 158 villages spread throughout most of the country.

There are some advantages to using nationally representative large sample surveys such as this. The main advantage is the ability to draw broad conclusions that are nationally significant. The general applicability of conclusions from such analysis provides policy makers with general guidance regarding the direction of national policy. For example, the measurably higher household welfare attributable to FUGs in Nepal may justify continued policy support of community-based forestry management.

The broadly applicable conclusions about the whole country available from a national sample survey come with some costs. In particular, in the case of Nepal, the Living Standards Survey did not include sufficiently detailed information about household participation in FUGs, and it was deficient in its measurement of natural resource stocks, such as the quantity and quality of forest resources available to the households.

One solution to this problem may be to include an environment module in living standards surveys. This solution may not be always practicable, however. For example, the size of the survey instrument may prevent addition of an extra module. Household and community questionnaires may not be the best instruments with which to collect natural resource stock data.

A second solution is to augment national standard of living survey data with environmental data from other sources. The chief obstacle to this method is the absence of sufficient means of combining the two data sets at the appropriate level of aggregation. For example, the biomass stock at the community level may be one of the main determinants of participation in FUGs in Nepal. However, information on area under different types of forests was available only at the district level. Such mismatches between data sets hinder analysis. In another case study, in Malawi, remote sensing data on biomass stock were matched with living standard data at the community level. This was possible because latitude and longitude coordinate information were available in the household survey data set, permitting the matching of households with the location of forested areas (Bandyopadhyay, Shyamsundar, and Baccini 2006).

A third approach is to design and implement a specialized household survey to measure the welfare impact of a specific environmental policy change. This approach was undertaken in the case studies of land tenure in China, communal conservancies in Namibia, and IMTs in the Philippines. Budgetary considerations may restrict the scope and scale of such surveys, as was the case in China, Namibia, and the Philippines. The smaller scale and narrower scope of specialized surveys allow for much more detailed investigation of specific environmental and natural resource management issues and associated policy measures that may affect household welfare. In China the study was restricted to three provinces, Guizhou, Hunan, and Yunnan. In Namibia the survey was restricted to two regions and seven communal conservancies. In the Philippines the study looked at a single irrigation system.

A fourth source of data on the environment-poverty nexus may come from pilot projects, as was the case in Nicaragua. Monitoring and evaluation is an integral part of pilot projects. Such projects not only allow for tailored survey design and other methods of data collection, they may also allow for embedding tools for impact evaluation at the project design stage. For example, elements of randomized experiments may be included in the design of pilot projects to allow for better use of impact evaluation analysis methods. The Nicaragua case study was deficient in its selection of randomized control and treatment groups. The control group selected during the data collection stage was later determined to be different from the treatment households.

The availability of time-series data from Argentina allowed this study to use the difference-in-difference method of impact analysis. This estimate of health welfare impact does not suffer from the selection biases from observed and unobservable factors. This study is an example of one of the most-reliable methods of parametric impact estimation.

Randomized experiments, before and after data, and appropriate treatment and control groups are necessary to assign causality between changes in environmental resource management and household economic and health-based welfare measures. When a randomized experiment is not practical, successful impact analysis requires that researchers carefully choose appropriate control groups and collect data on relevant indicators of changes, observable selection factors, and outcomes.

Conclusion

The pathways between environmental policy reform and household welfare are varied and complex. One type of environmental reform is devolution of environmental and natural resource management to communities and private entities. Recent policy changes by many governments have allowed devolution of control and management of environmental resources to communities and private entities. PES programs in many countries provide direct economic incentives to households that engage in better environmental management of private natural resources. The case studies presented in this chapter illustrate how impact evaluation methods can be applied to household survey data to estimate quantitative associations between community-based environmental resource management and household welfare.

Five key messages emerge from the studies reviewed in this chapter:
1. Household participation in community-based management of environmental resources has had mixed results. Some studies show that participants derive larger welfare benefits than do nonparticipants. Other studies indicate that participating and nonparticipating households share benefits more equally.

2. In these studies, community-based environmental resource management has a positive and measurable impact on household welfare. Higher welfare stems from increased economic activities, reinvestment in community infrastructure, and effective management of resources.

3. The poor benefit more from most of the reform programs examined in this chapter. However, in two case studies, the landless (Bandyopadhyay, Shyamsundar, and Kanel 2006) and the extremely poor (Pagiola, Rios, and Arcenas forthcoming) do not benefit more than their richer counterparts. Measuring the distribution of benefits from policy reforms can confirm whether or not vulnerable groups receive the benefits, paving the way for better targeting in the future.

4. Measurement of the welfare impact of environmental reforms using data from randomized social experiments or data from before and after the reform is most desirable. However, such estimations are not always practical. Future analysis may benefit from more attention to control and treatment groups, before and after data collection, and randomized experiments where feasible.

5. Cross-sectional household data have limitations regarding establishing causality between environmental reforms and poverty alleviation. With appropriate treatment and control groups and selection of the right analytical tools, it is, however, possible to draw policy-relevant conclusions from cross-sectional household data.

Notes

1 Communal land refers to areas in which property is commonly held and some form of traditional authority is in place. In Namibia all communal land belongs to the state.

2 Best practice in any econometric exercise is to use base modeling and hypothesis tests on qualitative information about the context of the policy reforms on the ground. Quantitative data do not always capture specific aspects of implementation of each policy reform or local customs and conditions. Collection of qualitative information is vital to understanding and interpreting the quantitative data collected at the household level.

CHAPTER 5

Directions for Change

POOR HOUSEHOLDS HAVE LIMITED ASSETS they can use to make investments. They face fewer income-earning opportunities, are exposed to higher health risks, and are less able to cope with adverse economic and health shocks. In this context, it is appropriate to worry about environmental problems facing the poor and ask whether there is a way to reduce poverty through environmental management. The review of the analytical work in this area, while it raises some doubt about certain linkages between poverty and the environment, provides evidence of mechanisms that can lead to poverty reduction.

Use of Local Natural Resources

Resources serve as a significant source of income for many rural households. The case studies suggest that resource use may increase and dependence decrease with income. There is also some evidence that suggests a more nuanced picture: households that are neither the poorest nor the least poor may be the main beneficiaries of nature's bounty. This is possible because resources found in commons often complement private assets, such as land and livestock. The poorest, who lack these private resources, may be dependent on forests for energy and housing needs but less so for other purposes.

Access to resources can serve as a buffer or insurance during times of need. In poor countries with limited financial and credit markets, the poor may depend on friends and family, as well as commonly available resources during times of stress. The empirical evidence of household responses to unexpected shocks and the insurance role of natural resources is limited, however. This role needs to be better understood, through careful empirical studies.

Is local degradation likely to decrease in the near future, particularly if household wealth and income rise? Empirical evidence does not support this oft-made assumption. Local resource use is unlikely to dramatically decline in rural areas. One reason why households continue to degrade natural resources is that the impact of slow and small changes in resource availability on welfare is small. Households adapt to changes in resource availability by, for example, using alternate resources or obtaining their resources from alternate areas. As long as the opportunity cost of time is low, the welfare impact of degradation is likely to be small. Thus, better environmental management, increases in nonfarm and nonresource-based economic opportunities, and changes in regulatory policies are likely to be important in stemming degradation.

Both the poor and the nonpoor contribute to environmental loss. The lack of markets in some cases and growth in markets in others; poor governance institutions; high discount rates; and population growth all play roles. Many of the forces that contribute to significant changes in ecosystems originate from macroeconomic and policy changes that may have little to do with natural resource sectors. Reducing poverty among resource-dependent households may thus not directly or immediately contribute to improvements in local natural resource use. There is no substitute for environmental management as a component of a practical and strong regulatory framework to ensure sustainability.

Fisheries, lakes, animal populations, and various natural processes are able to withstand changes to a certain extent, but they may collapse if perturbed beyond natural thresholds—with significant negative impacts on the resource-dependent poor. Furthermore, the more the poor consume natural resources, the less will be available for the future, which may impoverish them further.

Is there evidence of such poverty traps, of a downward spiral of natural resource loss and increased poverty? While work within and outside the World Bank suggests that that this type of negative dynamic relationship may be present in some areas, this issue needs a great deal more examination. More research is needed to understand the complex dynamics of natural systems and the interlinkages to poor resource-dependent households. More work is also needed in examining environmental services, including flood control services and the hydrological functions of forests in aiding the poor.

Design Principles for Improving Environmental Health

Good environmental quality—particularly of air, water, and sanitation—is a necessary condition for improving the welfare of the poor. Empirical studies reinforce the message that health-focused policies and public investments should be based on much broader considerations, such as the environment, education, nutrition, and public health information. How to design and implement such multidimensional programs remains a challenge. The next step is to identify design principles that will allow for successful implementation of these more complex projects.

One component of the design of health-focused projects is increased emphasis on environmental infrastructure. Investments in clean water and sanitation infrastructure have external effects on household health. Publicly funded programs need to recognize and capture this externality. Targeting significant coverage of water and sanitation needs to be a key component of any "total sanitation" program.

More holistic projects also require greater emphasis on public health information. Lack of information about the health impact of poor air and water can affect demand for environmental quality and mitigating behavior by households. Some evidence suggests that health information can lead to behavioral responses that mitigate the adverse health effects of poor environmental conditions more than increases in wealth or improvements in education. Households do respond to information, particularly with regard to health issues; projects need to take this into account.

While some aspects of designing water and sanitation projects are fairly well known, there is a huge gap in the understanding of indoor air pollution and mechanisms to reduce its impacts. It is estimated that about 20 percent of the estimated 12 million annual deaths of children under five and about 10 percent of perinatal deaths are directly related to acute lower respiratory infection as a result of exposure to indoor air pollution (WHO 2002). As these numbers show, this is not a trivial problem. The question is what to do about it. The understanding of key factors that contribute to and reduce the impacts of indoor air pollution is limited. Many variables matter: energy technology, housing characteristics, and behavioral responses can all play roles. Should the focus be on promoting efficient wood stoves or transitioning fuel use from biomass to charcoal or kerosene? Is there a role for increased household information on housing structure and ventilation? Studies are needed that identify the relative importance of different factors that affect pollution and responses to pollution.

Better Data for Monitoring Change

How important is data collection and analysis in this area? Countries make huge investments in health and natural resource management projects. While many tools are available for assessing the success of these projects, it is hard to evaluate their poverty outcomes without quantitative data. Every policy change or investment need not be subject to a careful quantitative evaluation. However, a small but systematic effort to collect data and analyze outcomes would be very useful for making progress in this complex field.

The cross-sectional household data that are generally available for poverty-environment analyses are limited in their ability to establish causality. It is possible, however, to draw policy-relevant conclusions from these data through careful selection of analytical tools. Many of these conclusions will need to be qualified, and policy recommendations will need to account for the uncertainties involved.

One way forward is to consider "add-ons" to the Living Standards Measurement Surveys many countries conduct. Specific modules could be created to collect a subset of environmental health and natural resource management information. Including these modules only in specific surveys and specific countries would ensure the collection of longitudinal data that is vital to evaluate changes.

Future analysis would benefit tremendously from more attention to data collection in four areas: longitudinal studies, control and treatment comparisons, before and after intervention studies, and randomized experiments. In environmental health, more information is needed in specific areas, such as cause-of-death information and other retrospective information on social, environmental, and health conditions at the household level. Quantitative studies need to be complemented with in-depth and more contextual qualitative methods of analysis. Future policy analysis should aim to combine quantitative with qualitative approaches in order to provide more credible evidence for guiding the design and implementation of programs.

Policy Reforms for Managing the Environment and Reducing Poverty

The past two decades have seen new reforms in environmental management that have community participation and economic development as core goals. The studies reviewed in this report focus on reforms that strengthened community rights, created stronger incentives for resource management, and developed new markets that facilitated payments for environmental services. The report also examined reforms outside the environment sector that strengthened private property rights and increased access to services.

A key conclusion from these studies is that decentralization of natural resource management is beginning to work in some communities. Although it does not work as well as it should and there are many layers of challenges, community-based resource management can have a positive and measurable impact on household welfare. This result does not, of course, hold true for all examples of decentralized resource management.

The improved benefits from community management of local resources appear to come from three sources: investment in community infrastructure, increased economic activities, and effective management of resources. All of these aspects of community projects need additional support and additional monitoring.

An important question raised about community-oriented resource management programs is whether participation is equitable or captured by the elite. The studies reviewed here suggest that participation is not always limited to community elites, and the welfare benefits of participation in such programs are not always greater for participating households. There are diverse distributional impacts of community-based natural resource management programs. In two cases, the landless and the extremely poor do not benefit any more than their richer counterparts. There is scope for investigating how local political and power positions determine who participates and how economic profits are allocated among households. Researchers should expect to be surprised; some standard hypotheses may not hold.

The poor are willing to participate in fairly complex environmental management programs if these programs provide the right incentives. Emerging evidence also suggests that the poor are willing to contribute to the provision of environmental services. In Nicaragua, for example, poor households were willing to implement changes that brought about public benefits such as increased biodiversity and higher carbon sequestration in return for payments for these services.

Two other case studies focus on strengthening incentives through irrigation management and land reform. Both studies suggest that there are positive productivity benefits from such reforms. Do these reforms, which often stem from nonenvironmental considerations, strengthen sustainable resource use? Further examination of the physical changes brought about would help identify long-term impacts on sustainability.

Increasing access to environmental infrastructure for safe water and sanitation can decrease child mortality. The evidence for this from Argentina is particularly strong, because the analytical methods employed eliminated selection bias and reduced the potential impact of unobserved variables.

Moving Forward

Poverty reduction and sustainable resource use go hand in hand under certain circumstances and not in others. Going forward, policymakers may benefit from the following insights:

- The poor are dependent on local resources for income and consumption. Not enough is known about the role of commons in providing a buffer or insurance; the dynamics of ecosystem changes and their impacts on the poor; or the value of various natural services, particularly related to mitigating natural disasters. Careful analyses are needed in these areas.

- Mechanisms to reduce indoor air pollution are not very well understood. Improving the quality of indoor air will affect health and possibly forest use, with potential implications for carbon sequestration. It would be particularly useful to design joint "intervention and analyses" projects on this issue.

- Health programs need to pay more attention to both the coverage of interventions (to capture positive externalities) and the role of health information in prompting behavioral change. A broad-based approach toward health should be adopted that embraces environmental as well as more traditional health-sector interventions.

- The poor are willing to participate in a variety of resource management programs, some of which lead to significant welfare improvements. Prudent investments need to continue to be made in projects that create new incentives and strengthen property rights.

Ensuring that environmental management projects help the poor is an important and continuous challenge. Equally important is the complementary task of ensuring that poverty reduction programs contribute to sustainable development. Increased efforts are needed to collect good quantitative and qualitative data to help monitor and evaluate these programs.

References

Adhikari, B. 2003. "Property Rights and Natural Resources: Socioeconomic Heterogeneity and Distributional Implications of Community Forest Management." SANDEE Working Paper 1-03, Katmandu.

———. 2005. "Poverty, Property Rights and Collective Action: Understanding the Distributive Aspects of Common Property Resource Management." *Environment and Development Economics* 10 (1): 7–31.

Agarwal, B. 2001. "Participatory Exclusions, Community Forestry, and Gender: An Analysis for South Asia and a Conceptual Framework." *World Development* 29 (10): 1623–48.

Ahmed, Kulsum, Yewande Awe, Maureen Cropper, Douglas Barnes, and Masami Kojima. 2005. *Environmental Health and Traditional Fuel Use in Guatemala.* World Bank, Energy Sector Management Assistance Program, Washington, DC.

Alderman, H., and M. Garcia. 1994. "Food Security and Health Security: Explaining the Levels of Nutritional Status in Pakistan." *Economic Development and Cultural Change* 42 (3): 485–508.

Alderman, H., J. Hentschel, and R. Sabates. 2003. "With the Health of Ones' Neighbors: Externalities in the Production of Nutrition in Peru." *Social Science and Medicine* 56 (10): 2019–31.

Angelsen, A., and D. Kaimowitz. 2001. "Agricultural Technology and Forests: A Recapitulation." In *Agricultural Technologies and Tropical Deforestation*, ed. A. Angelsen and D. Kaimowitz. Wallingford, United Kingdom: CABI Publishing.

Bagadion, B. 2002. "Role of Water Users Associations for Sustainable Irrigation Management." In *Organizational Change for Participatory Irrigation Management.* Report of the Asian

Productivity Organization Seminar on "Organizational Change for Participatory Irrigation Management," held in Manila, October 23–27. Tokyo: Asian Productivity Organization.

Baland, Jean-Marie, Pranab Bardhan, Sanghamitra Das, Dilip Mookherjee, and Rinki Sarkar. 2006. "Managing the Environmental Consequences of Growth: Forest Degradation in the Indian Mid-Himalayas." University of California, Berkeley.

Baland, Jean-Marie, and Patrick François. 2005. "Commons as Insurance and the Welfare Impact Of Privatization." *Journal of Public Economics* 80 (2–3): 211–31.

Bandyoadhyay, Sushenjit, Michael Humavindu, Priya Shyamsundar, and Limin Wang. 2004. "Do Households Gain from Community-Based Natural Resource Management? An Evaluation of Community Conservancy in Namibia." Policy Research Working Paper 3337, World Bank, Washington, DC.

Bandyopadhyay, Sushenjit, and Priya Shyamsundar. 2004. "Fuelwood Consumption and Participation in Community Forestry in India." Policy Research Working Paper 3331, World Bank, Washington, DC.

Bandyopadhyay, Sushenjit, Priya Shyamsundar, and Alessandro Baccini. 2006. "Forests Biomass Use, and Poverty in Malawi." Policy Research Working Paper 4068, World Bank, Washington, DC.

Bandyopadhyay, Sushenjit, Priya Shyamsundar, and Keshav Raj Kanel. 2006. "Forestry User Groups in Nepal: Can Institutional Change Lead to Economic Development?" Background paper for the Nepal Poverty Assessment, World Bank, Washington, DC.

Bandyopadhyay, Sushenjit, Priya Shyamsundar, and Mei Xie. 2007. "Yield Impact of Irrigation Management Transfer: A Success Story from the Philippines." Processed. Environment Department, World Bank.

Barrera, A., 1990. "The Role of Maternal Schooling and Its Interaction with Public Health Programs in Child Health Production." *Journal of Development Economics* 32 (1): 69–91.

Barrett, Christopher. 2004. "Rural Poverty Dynamics: Development Policy Implications." *Agricultural Economics* 32 (1): 43–58.

Barrett, Christopher, and Peter Arcese. 1998. "Wildlife Harvest in Integrated Conservation and Development Projects: Linking Harvest to Household Demand, Agricultural Production and Environmental Shocks in Serengeti." *Land Economics* 74 (4): 449–65.

Bhargrava, A. 2003, "Family Planning, Gender Differences and Infant Mortality: Evidence from Uttar Pradesh, India." *Journal of Econometrics* 112 (1): 225–40.

Bluffstone, R. 1995. "The Effects of Labor Markets on Deforestation in Developing Countries under Open Access: An Example from Rural Nepal." *Journal of Environmental Economics and Management* 29 (10): 42–63.

Boy, E., N. Bruce, and H. Delgado. 2002. "Birth Weight and Exposure to Kitchen Wood Smoke during Pregnancy in Rural Guatemala." *Environmental Health Perspectives* 110 (1): 109–114.

Brasselle, A. S., F. Gaspart, and J. P. Platteau. 2002. "Land Tenure Security and Investment Incentives: Puzzling Evidence from Burkina Faso." *Journal of Development Economics* 67 (2): 373–418.

Brown, Kenneth H. 2003. "Diarrhea and Malnutrition." *Journal of Nutrition* 133 (1): S328–S332.

Cairncross, Sandy, and Vivian Valdemanis. 2006. "Water Supply, Sanitation and Hygiene Promotion." In *Disease Control Priorities in Developing Countries,* 2nd ed. Dean T. Jamison, J. G. Breman, A. R. Measham, G. Alleyne, M. Claeson, D. B. Evans, P. Jha, A. Mills, and P. Musgrove. Washington, DC: Oxford University Press and the World Bank.

Cavendish, W. 2000. Empirical Regularities in the Poverty-Environment Relationship of Rural Households: Evidence from Zimbabwe. *World Development* 28 (11): 1979–2003.

Chapman, Robert S., Xingzhou He, Aaron E. Blair, and Qing Lan. 2005. "Improvement in Household Stoves and Risk of Chronic Obstructive Pulmonary Disease in Xuanwei, China: Retrospective Cohort Study." *British Medical Journal* 331 (7524): 1050.

Chettri-Khattri, Arun. 2007. "Who Pays for Conservation: Evidence from Forestry in Nepal." In *Promise, Trust and Evolution: Managing the Commons of South Asia*, ed. R. Ghate, N. C. Jodha, and P. Mukhopadhyay. Oxford: Oxford University Press.

Cole, T. J., and J. M. Parkin. 1977. "Infection and Its Effects on Growth of Young Children: A Comparison of the Gambia and Uganda." *Transaction of the Royal Society of Tropical Medicine and Hygiene* 71: 196–98.

Dasgupta, Partha. 2003. "Population, Poverty and the Natural Environment." In *Handbook of Environmental Economics*, vol. 1, ed. K-G. Maler and J. Vincent. Amsterdam: Elsevier Press.

———. 2004. "World Poverty: Causes and Pathways." In *World Bank Conference on Development Economics*, ed. B. Pleskovic and N. H. Stern. Washington, DC: World Bank.

Dasgupta, S., M. Huq, M. Khaliquzzaman, K. Pandey, and D. Wheeler. 2006a. "Indoor Air Quality for Poor Families: New Evidence from Bangladesh." *Indoor Air* 16 (6): 426–44.

———. 2006b. "Who Suffers from Indoor Air Pollution? Evidence from Bangladesh." *Health Policy and Planning* 21 (6): 444–58.

Deininger, Klaus, and Songqing Jin. 2003. "The Impact of Property Rights on Households' Investment, Risk Coping, and Policy Preferences: Evidence from China." *Economic Development and Cultural Change* 51 (4): 851–82.

Duraiappah, Anantha, K. 1998. "Poverty Environment Degradation: A Review and Analyses of the Nexus." *World Development* 26 (12): 2169–79.

Dzinggirai, V. 2003. "The New Scramble for African Countryside." *Development and Change* 34 (2): 243–63.

Edmonds, E. 2002. "Government-Initiated Community Resource Management and Local Resource Extraction from Nepal's Forests." *Journal of Development Economics* 68 (1): 89–115.

Esrey, Steve. 1996. "Water, Waste and Well-Being: A Multi-Country Study." *American Journal of Epidemiology* 143 (6): 608–23.

Ezzati, M., and D. M. Kammen. 2001. "Indoor Air Pollution from Biomass Combustion and Acute Respiratory Infections in Kenya: An Exposure-Response Study." *Lancet* 358 (9282): 619–24.

———. 2002. "The Health Impacts of Exposure to Indoor Air Pollution from Solid Fuels in Developing Countries: Knowledge, Gaps and Data Needs." *Environmental Health Perspectives* 110 (11): 1057–68.

Ezzati, M., B. N. Mbinda, and D. M. Kammen. 2000. "Comparison of Emission and Residential Exposure from Traditional and Improved Biofuel Stoves in Rural Kenya." *Environmental Science Technology* 34 (4): 578–83.

Ezzati, M., H. Saleh, and D. M. Kammen. 2000. "The Contributions of Emissions and Spatial Microenvironments to Exposure to Indoor Air Pollution from Biomass Combustion in Kenya." *Environmental Health Perspectives* 108 (9): 833–39.

Fewtrell, Lorna, and John M. Colford. 2004. "Water, Sanitation and Hygiene: Interventions and Diarrhoea: A Systematic Review and Meta-analysis." World Bank, Health, Nutrition and Population Discussion Paper, Water Supply and Sanitation Sector Board, Washington, DC.

Fisher, M., G. Shively, and S. Buccola. 2005. "Activity Choice, Labor Allocation and Forest Use in Malawi." *Land Economics* 81 (4): 503–17.

Galiani, S., P. Gertler, and E. Schargrodsky. 2005. "Water for Life: The Impact of the Privatization of Water Services on Child Mortality." *Journal of Political Economy* 113 (1): 83–120.

Gauderman, W. James, Edward Avol, Frank Gilliland, Hita Vora, Duncan Thomas, Kiros Berhane, Rob McConnell, Nino Kuenzli, Fred Lurmann, Edward Rappaport, Helene Margolis, David Bates, and John Peters. 2004. "The Effect of Air Pollution on Lung Development from 10 to 18 Years of Age." *New England Journal of Medicine* 351: 1057–67.

Glewwe, P. 1999. "Why Does Mothers' Schooling Raise Child Health in Developing Countries: Evidence from Morocco." *Journal of Human Resources* 34 (1): 124–36.

Godoy, R., D. Wilke, H. Overman, A. Cubas, G. Cubas, J. Demmer, K. McSweeney, and N. Brokaw. 2000. "Valuation of Consumption and Sale of Forest Goods from A Central American Rain Forest." *Nature* 406 (6): 62–63.

Gragnolati, M. 1999. "Child Malnutrition in Rural Guatemala: A Multilevel Statistical Analysis." Ph.D. dissertation, Princeton University, Princeton, NJ.

Groenfeldt, D., and M. Svendsen, eds. 2000. *Case Studies in Participatory Irrigation Management.* Washington, DC: World Bank Institute.

Hughes, G., K. Lvovsky, and M. Dunleavy. 2001. "Environmental Health in India: Evidence from Andhra Pradesh." World Bank, South Asia Region, Environment and Social Development Division, Washington, DC.

Imbens, Guido. 2004. "Nonparametric Estimation of Average Treatment Effects under Exogeneity: A Review." *Review of Economics and Statistics* 86 (1): 4–29.

Jacoby, H. G., Guo Li, and Scott Rozelle. 2003. "Hazards of Expropriation: Tenure Insecurity and Investments in Rural China." *American Economic Review* 92 (5): 1420–47.

Jacoby, H., and L. Wang. 2004. "Environmental Determinants of Child Mortality in Rural China: A Competing Risks Approach." Research Working Paper 3241, World Bank, Washington, DC.

Jalan, J., and M. Ravallion. 2003. "Does Piped Water Reduce Diarrhea for Children in Rural India?" *Journal of Econometrics* 112 (1): 153–73.

Jalan, J., and E. Somanathan. 2004. "The Importance of Being Informed: Experimental Evidence on the Demand for Environmental Quality." SANDEE Working Paper 8-04, South Asian Network for Development and Environmental Economics, Kathmandu.

Jodha, N. S. 1986. "Common Property Resources and the Rural Poor in Dry Regions of India." *Economic and Political Weekly* 21 (27): 1169–81.

Jones, B. 1999a. "Policy Lessons from the Evolution of a Community-Based Approach to Wildlife Management, Kunene Region, Namibia." *Journal of International Development* 11 (2): 295–304.

————. 1999b. "Rights Revenues and Resources: The Problems and Potentials of Conservancies as Community Wildlife Management Institutions in Namibia." Evaluating Eden Series Discussion Paper 2, International Institute for Environment and Development, London.

Jones, B., and M. Murphree. 2001. "The Evolution of Policy and Community Conservation in Namibia and Zimbabwe." In *African Widife and Livelihoods: The Promise and Performance of Community Conservation,* ed. D. Hulme and M. Murphree. Oxford: James Currey Ltd.

Kaimowitz, D., O. Erwidodo, P. Ndoye, P. Pacheco, P. Balanza, and W. D. Sunderlin. 1998. "Considering the Impact of Structural Adjustment Policies on Forests in Bolivia, Cameroon and Indonesia." *Unasylva* 49 (194): 57–64.

Kanel, K. R. 2004. "Twenty-Five Years of Community Forestry: Contribution to Millennium Development Goals." In *Proceedings of the Fourth National Workshop on Community Forestry*. Community Forestry Division, Department of Forests, Kathmandu.

Keane, Michael. 2006. "Structural vs. Atheoretical Approaches to Econometrics." Department of Economics, Yale University, New Haven, CT.

Landell-Mills, N., and I. Porras. 2002. *Silver Bullet or Fools' Gold? A Global Review of Markets for Forest Environmental Services and Their Impact on the Poor*. International Institute for Environment and Development, London.

Lavy, V., J. Strauss, D. Thomas, and P. De Vreyer. 1996. "Quality of Health Care, Survival and Health Outcomes in Ghana." *Journal of Health Economics* 15 (3): 333–57.

Lee, L. F., M. R. Rosenzweig, and M. M. Pitt. 1992. "The Effects of Nutrition, Sanitation, and Water Purity on Child Health in High Mortality Populations." *Journal of Econometrics* 77 (1): 209–35.

Lin, J. Y. 1992. "Rural Reforms and Agricultural Growth in China." *American Economic Review* 82 (1): 34–51.

Luby, Stephen, Mubina Agboatwalla, Daniel Feikin, John Painter, Ward Billhimer, and Robert Hoekstra. 2005. "Effect of Hand Washing on Child Health: A Randomized Controlled Trial." *Lancet* 366 (9481): 225–33.

Lybbert, Travis, J., Christopher B. Barrett, Solomon Desta, and D. Layne Coppock. 2004. "Stochastic Wealth Dynamics and Risk Management among a Poor Population." *Economic Journal* 114 (498): 750–77.

Lybbert, Travis J., Christopher B. Barret, and Hamid Narjisse. 2002. "Market-Based Conservation and Local Benefits: The Case of Argan Oil in Morocco." *Ecological Economics* 41 (1): 125–44.

Mata, L. 1978. *The Children of Santa Maria Cauque: A Prospective Field Study of Health and Growth*. Cambridge, MA: MIT Press.

McSweeney, K. 2005. "Natural Insurance, Forest Access and Compounded Misfortune: Forest Resources in Smallholder Coping Strategies before and after Hurricane Mitch, North Eastern Honduras." *World Development* 33 (9): 1453–71.

Merrick, T. 1985. "The Effect of Piped Water on Early Childhood Mortality in Urban Brazil, 1970 to 1976." *Demography* 22 (1): 1–24.

Moffitt, Robert. 2003. "The Role of Randomized Field Trials in Social Science Research: A Perspective from Evaluations of Reforms of Social Welfare Programs." CeMMAP Working Paper CWP23/02, Department of Economics, University College, London.

Mosley, W., and L. C. Chen. 1984. "An Analytical Framework for the Study of Child Survival in Developing Countries." *Population and Development Review* 10 (Supplement): 25–45.

Mulreany, J. P., S. Calikoglu, S. Ruiz, and J. W. Sapsin. 2006. "Water Privatization and Public Health in Latin America." *Pan-American Journal of Public Health* 19 (1): 23–32.

Munyao, K., and C. Barrett. 2006. "Decentralization of Pastoral Resources Management and Its Effect on Environmental Degradation and Poverty: Experience from Northern Kenya." Department of Applied Economics and Management, Cornell University, Ithaca, NY.

NACSO (Namibian Association of CBNRM Support Organization). 2005. *Namibia's Communal Conservancies: An Overview of Status, Progress and Potential of Namibia's Communal Area Conservancies 2004*. Windhoek.

Narain, Urvashi, Klaas vant Veld, and Shreekant Gupta. 2005. "Poverty and the Environment: Exploring the Relationship between Household Incomes, Private Assets, and Natural Assets." Resources for the Future Discussion Paper 05-18, Washington, DC.

Pagiola, S., A. Arcenas, and G. Platais. 2005. "Can Payments for Environmental Services Help Reduce Poverty? An Exploration of the Issues and the Evidence to Date from Latin America." *World Development* 33 (2): 237–53.

Pagiola, S., N. Landell-Mills, and J. Bishop. 2002. "Making Market-Based Mechanisms Work for Forests and People." In *Selling Forest Environmental Services: Market-Based Mechanisms for Conservation and Development*, ed. S. Pagiola, J. Bishop, and N. Landell-Mills. London: Earthscan.

Pagiola, Stefano, Ana R. Rios, and Agustin Arcenas. Forthcoming. "Can the Poor Participate in Payments for Environmental Services? Lessons from Silvopastoral Project in Nicaragua." *Environment and Development Economics.*

Pattanayak, Subhrendu, and Erin O. Sills. 2001. "Do Tropical Forests Provide Natural Insurance? The Microeconomics of Non-Timber Forest Product Collection in the Brazilian Amazon." *Land Economics* 77 (4): 595–612.

Pinstrup-Andersen, P., D. Pelletier, and H. Alderman, eds. 1995. *Child Growth and Nutrition in Developing Countries: Priorities for Action.* Ithaca, NY: Cornell University Press.

Pitt, M., M. Rosenzweig, and M. N. Hassan. 2006. "Sharing the Burden of Disease: Gender, the Household Division of Labor and the Health Effects of Indoor Air Pollution." CID Working Paper 119, Center for International Development, Harvard University, Cambridge, MA.

Pruss-Ustun, Annette, David Kay, Lorna Fewtrell, and Jamie Bartram. 2004. "Unsafe Water, Sanitation and Hygiene." In *Comparative Quantification of Health Risks: Global and Regional Burden of Disease due to Selected Major Risk Factors*, ed. M. Ezzati, A. Lopez, A. Rodgers, and C. Murray. Geneva: World Health Organization.

Qi, J. C. 1999. "Two Decades of Rural Reform in China: An Overview and Assessment." *China Quarterly* 159: 616–28.

Quick, R. E., A. Kimura, A. Thevos, M. Tmbo, I. Shamputa, L. Hutwagner, and E. Mintz. 2002. "Diarrhea Prevention through Household-Level Water Disinfection and Safe Storage in Zambia." *American Journal of Tropical Medicine and Hygiene* 66 (5): 584–89.

Quick, R., L. Venczel, E. Mintz, L. Soleto, J. Aparicio, M. Gironaz, L. Hutwagner, K. Greene, C. Bopp, K. Maloney, D. Chavez, M. Sobsey, and R. Tauxe. 1999. "Diarrhea Prevention in Bolivia through Point-of-Use Disinfection and Safe Storage: A Promising New Strategy." *Epidemiology and Infection* 122 (1): 83–90.

Ravallion, M. 2007. "Evaluating Anti-Poverty Programs." *Handbook of Development Economics*, vol. 4, ed. T. P. Schultz and J. Strauss. Amsterdam: North-Holland.

Reardon, T., and S. A. Vosti. 1995. "Links between Rural Poverty and the Environment in Developing Countries: Asset Categories and Investment Poverty." *World Development* 23 (9): 1495–1506.

Ridder, G., and I. Tunali. 1999. "Stratified Partial Likelihood Estimation." *Journal of Econometrics* 92 (2): 193–232.

Rutstein, S., and K. Johnson. 2004. "The DHS Wealth Index." DHS Comparative Report 6, ORC Macro, Calverton, MD.

Sabio, E. A., and A. D. Mendoza. 2002. "Philippines." In *Organizational Change for Participatory Irrigation Management*. Report of the Asian Productivity Organization Seminar on "Organizational Change for Participatory Irrigation Management," held in Manila, October 23–27. Tokyo: Asian Productivity Organization.

Sarin, M., with L. Ray, M.S. Raju, M. Chatterjee, N. Banerjee, and S. Hiremath. 1988. *Who Gains? Who Loses? Gender and Equity Concerns in Joint Forest Management.* Society for Promotion of Wasteland Development, New Delhi.

Semenza, J. C., L. Roberts, A. Henderson, J. Bogen, and C. H. Rubin. 1998. "Water Distribution System and Diarrheal Disease Transmission: A Case Study in Uzbekistan." *American Journal of Tropical Medicine and Hygiene* 59 (6): 941–46.

Shah, T., B. Koppen, D. Merrey, M. Lange, and M. Samad. 2002. "Institutional Alternatives in African Smallholder Irrigation: Lessons from International Experience with IMT." Research Report 60, International Water Management Institute, Colombo, Sri Lanka.

Shyamsundar, P., E. Araral, and S. Weeraratne. 2005. "Devolution of Resource Rights, Poverty, and Natural Resource Management: A Review." Environment Department Paper 104, World Bank, Washington, DC.

Siela and Danida. 2005. "A National Opinion Poll: Commune Council's Perception of Its Natural Resource Base and Livelihood Options." In *Poverty-Environment Nexus: Sustainable Approaches to Poverty Reduction in Cambodia, Laos, and Vietnam.* Washington, DC: World Bank.

Silva, P. 2005a. "Environmental Factors and Children's Malnutrition in Ethiopia." Research Working Paper 3489, World Bank, Washington, DC.

———. 2005b. "Exploring the Link between Poverty, Marine Protected Area Management and the Use of Destructive Fishing Gear." Policy Research Working Paper 3831, World Bank, Washington DC.

Sjaastad, Espen, Arild Agelsen, Pal Vedeld, and Jan Bojo. 2005. "What Is Environmental Income?" *Ecological Economics* 55 (1): 37–46.

Smith, Kirk. 2006. "Presentations from the Guatemala Air Pollution Intervention Trial (RESPIRE)." http://ehs.sph.berkeley.edu/krsmith/page.asp?id=19

Smith, Kirk R., S. Mehta, M. Maeusezahl-Feuz. 2004. "Indoor Smoke from Household Solid Fuels." In *Comparative Quantification of Health Risks: Global and Regional Burden of Disease due to Selected Major Risk Factors,* vol. 2, ed. M. Ezzati, A. D. Rodgers, A. D. Lopez, and C. J. L. Murray, 1435–93. Geneva: World Health Organization.

Smith, Kirk R., Jonathan M. Samet, Isabelle Romieu, and Nigel Bruce. 2000. "Indoor Air Pollution in Developing Countries and Acute Lower Respiratory Infections in Children." *Thorax* 55 (6): 518–32.

Smith, L., and L. Haddad. 1999. "Explaining Child Malnutrition in Developing Countries: A Cross-Country Analysis." Food Consumption and Nutrition Division Discussion Paper, International Food Policy Research Institute, Washington, DC.

Sundar, N. 2000. "Unpacking the 'Joint' in Joint Forest Management." *Development and Change* 31 (1): 255–79.

Sunderlin, W. D., A. Angelsen, B. Belcher, P. Burgers, R. Nasi, L. Santoso, and S. Wunder. 2005. "Livelihoods, Forests and Conservation in Developing Countries: An Overview." *World Development* 33 (9): 1383–1402.

Sunderlin, W. D., I. A. P. Resosudarmo, E. Rianto, and A. Angelsen. 2000. "The Effect of Indonesia's Economic Crisis on Small Farmers and Natural Forest Cover in the Outer Islands." CIFOR Occasional Paper 28 (1), Center for International Forestry Research, Bogor, Indonesia.

Tachibana, T., H. K. Upadhyay, R. Pokharel, S. Rayamajhi, and K. Otsuka. 2001. "Common Property and Forest Management in the Hill Region of Nepal." In *Land Tenure and Natural Resource Management: A Comparative Study of Agrarian Communities in Asia and Africa,* ed. K. Otsuka and F. Place. Baltimore, MD: Johns Hopkins University Press for the International Food Policy Research Institute.

Takasaki, Yoshita, Bradford L. Barham, and Oliver T. Coomes. 2004. "Risk-Coping Strategies in Tropical Forests: Flood, Illnesses and Resource Extraction." *Environment and Development Economics* 9 (2): 203–224.

Thomas, D., V. Lavy, and J. Strauss. 1996. "Public Policy and Anthropometric Outcomes in Côte d'Ivoire." *Journal of Public Economics* 61 (2): 155–92.

Van der Klaauw, B., and L. Wang. 2005. "Child Mortality in Rural India." Research Working Paper WPS3281, World Bank Washington, DC.

Vedeld, Paul, Arild Angelsen, Espen Sjaastad, and Gertrude K. Berg. 2004. "Counting on the Environment: Forest Incomes and the Rural Poor." Environment Department Paper 98, World Bank, Washington, DC.

Vermillion, D. L. 1992. "Irrigation Management Turnover: Structural Adjustment or Strategic Evolution?" *IIMI Review* 6 (2): 3–12.

————. 1997. "Impacts of Irrigation Management Transfer: A Review of the Evidence." Research Report 11, International Irrigation Management Institute, Colombo, Sri Lanka.

Wang, L. 2003. "Determinants of Child Mortality in LDCs: Empirical Findings from Demographic and Health Surveys." *Health Policy* 65 (3): 277–99.

WHO (World Health Organization). 2002. *The World Health Report 2002: Reducing Risks, Promoting Healthy Life.* Geneva: WHO.

Wolpin, K. I. 1997. "Determinants and Consequences of the Mortality and Health of Infants and Children." In *Handbook of Population and Family Economics*, ed. M. R. Rosenzweig and O. Stark, vol 1A. Amsterdam: Elsevier.

Wooldridge, Jeffrey. 2002. *Econometric Analysis of Cross-Section and Panel Data.* Cambridge, MA: MIT Press.

World Bank. 2001. "Health and Environment." Background paper for the World Bank Environment Strategy, Washington, DC.

————. 2002. *India: Household Energy, Indoor Air Pollution, and Health.* South Asia Environment and Social Development Unit, Washington, DC.

————. 2005. *Lessons Learned from Bangladesh, India, and Pakistan: Scaling-Up Rural Sanitation in South Asia.* Water and Sanitation Program Report, Washington, DC.

————. 2006a. *At Loggerheads? Agricultural Expansion, Poverty Reduction and Environment in the Tropical Forests.* Washington, DC: World Bank.

————. 2006b. *Poverty-Environment Nexus: Sustainable Approaches to Poverty Reduction in Cambodia, Lao PDR, and Vietnam.* Environment and Social Development Department, East Asia and the Pacific Region, Washington, DC.

————. 2006c. *Repositioning Nutrition as Central to Development: A Strategy for Large-Scale Action.* Directions in Development Series. Washington, DC: World Bank.

————. 2006d. *World Development Indicators.* Washington, DC: World Bank.

————. 2006e. *Where Is the Wealth of Nations? Measuring Capital for the 21st Century.* Washington, DC: World Bank.

Wunder, Sven. 2001. "Poverty Alleviation and Tropical Forests: What Scope for Synergies?" *World Development* 29 (1): 1817–33.

————. 2005. "Payments for Environmental Services: Some Nuts and Bolts." CIFOR Occasional Paper 42, Center for International Forestry Research, Bogor, Indonesia.

Zbinden, S., and D. Lee. 2005. "Paying for Environmental Services: An Analysis of Participation in Costa Rica's PSA program." *World Development* 33 (2): 255–72.

INDEX

Boxes, figures, notes, and tables are indicated by b, f, n, and t, respectively.

A

access rules, 14
analytical framework, 29, 31
analytical tools, 10
Argentina, 42, 51*b*, 57
At Loggerheads, 23

B

bias, 31, 32, 48*b*
biomass fuel, 9–10, 20, 21*f*
   density increase effects on poor, 22*f*
   indoor air pollution, 39, 43

C

Cambodia, 12b–13*b*
case-control studies, 43*n*
causality analysis, 10, 11–12
child health, 29, 30*t*
   indoor air pollution exposure, 40
   malnutrition, 4, 29, 36–38
child mortality, 30*t*, 41–42
   sanitation access, 35
   U-5 (*See* U-5 mortality)
   water access, 34, 38*b*, 42
China, 53–54
   study on child mortality, 41–42
   water access effects on children, 33, 34
climate change, 6
common property, 15–18, 16, 47
   income, 13–14
common property rights, 47–50, 52–53
   devolution of, 52
Communal Areas Management Program
   for Indigenous Resources
   (CAMPFIRE), 46
communal land, 60*n*
community participation, 47

community resource management, 7, 50*b*, 59,
   64–65
   beneficiaries of, 51*b*
   environmental income, 14
conservancies, 47, 49, 52
conservation programs, meat
   distribution, 16–17
conservation, financing of, 55
consumption smoothing, 22*f*
cooking fuels, 31, 35

D

data, 58, 60*n*
   collection, 43, 63–64
   limitations, 42
      cross-sectional, 49*b*, 60, 64
      household surveys, 57–59
decetralization of resource management,
   8, 64
deforestation, 23
degradation, 9, 62
   and household welfare, 20, 22, 25
   linkages to poverty, 11–12, 23–24
diarrheal disease, 34, 35–36, 37
disease burden, 27, 39, 44*n*
dose-response functions, 39

E

econometric techniques, 32
economic growth, 18–19, 24
electricity access, 34–35
elite capture, 47, 51*b*, 65
Enterprise Works, 44*n*
environment, 2–3, 27–28
   impact on health outcomes, 9–10
   linkage to poverty at household level, 5*f*
   poverty role, 22–24
   role in health outcomes, 8

environmental contaminants, 31–32
    child health outcomes, 29, 30t, 31
environmental health, 28, 63, 64
environmental income, 9, 24, 25n
    contributions to poor, 12–15
    resource poor and rich areas, 15t
environmental indicators, 4–5, 4f
environmental infrastructure, 3, 3t, 42, 63
environmental infrastructure investment, 53
environmental loss, 23, 62. *See also*
    degradation
environmental management, 1, 16–18
    and household welfare, 5–7
    and poverty reduction, 6t, 9, 25, 64–65
    community-based, 50b, 59–60
    PES, 8
error-in-variables problem, 32

F

fisheries' decline, 13b
fishing gear and practices and degradation, 23
forest management, 47, 50b, 52, 53
forest scarcity, 20
forest user groups (FUGs), 47, 52, 53, 57
forests, 2–3, 19, 23
    decline, 13b
    household income, 15, 19, 20
    individual access to, 52
fuelwood use, 19, 22, 50b, 52, 53

H

hand washing, 36
health information dissemination, 36b, 63
health linkages to environment, 28
health outcomes, 3, 3t, 8, 28–29
    child malnutrition, 36–38
    child morbidity, 35–36
    child mortality, 33–35
    children, 29, 30t, 31
    due to environmental risk exposure, 41
health programs, 66
health risk factors, 4, 27
health services, access to, 30t
height by age, 38
Honduras, 16, 17b, 17bt
household budget allocation, 44n
household income, 10n, 15. *See also*
    environmental income
household survey, design of, 58
household surveys, data limitations,
    31, 41, 42, 57, 64
household welfare, 20, 22, 52, 63
    adapting to resource availability, 62

and environmental management, 5–7
fuel use, 22–23
impact of community management, 60
increase due to community forestry, 53
targeting investments, 15
households, rural, 11, 61
Hurricane Mitch, 16, 17b, 17bt

I

impact estimation, 50b
impact evaluation methods, 48b–49b, 59
income. *See* environmental income; household
    income; natural resources
India, 14, 19, 50b
    child mortality, 41–42
indoor air pollution, 35, 43, 44n, 63, 66
    health risks, 39–40
information dissemination, 36b, 63
information gap, 8
information role in policy, 9
infrastructure. *See* environmental infrastructure
investments to improve households, 15
irrigation management transfers (IMTs), 54, 55

J

Jhabua, Madhya Pradesh, India, 14

K

Kunene, Namibia, 52

L

land rights, 46
land tenure, 54
Living Standards Measurement Surveys
    (LSMS), 37, 57, 58, 64

M

Magat River Integrated Irrigation System
    (MRIIS), 55
Malawi, biomass use, 20, 21f, 22f
malnutrition, 4, 27, 37, 44n
    children, 29, 30t, 36–38
management incentives, 54–55
maternal education, 37
meat distribution program, 16–17
Millennium Development Goals (MDGs), 4

N

Namibia, common property rights, 47–50, 52
natural resource availability, effects on
    poverty, 20

natural resource management, 6, 8, 64
  community-based, 7, 60, 64–65
natural resources, 10, 16, 45–46, 62
  and poverty and U-5 mortality, 2t
  buffering function, 16–18
  demand due to growth, 18–19
  dependence on, 2, 7, 11, 12, 12b–13b,
    13bf, 14, 19, 65
  depletion trends, 13b
  income for rural households, 61
  linked with poverty and environment, 4–5
  sustainability, 65–66
Nepal, 14, 51b
  common property rights, 52–53
  study on wealth and fuelwood
    use, 19, 57
Nicaragua, 51b, 55–56, 65
nutrition, child, 30t
nutritional status studies, 42

P

Pacaya-Samiria National Reserve, 22
particulate matter exposure, 39
payments for environmental services (PES),
  8, 51b, 55–56, 59
Peru, 37
Philippines, the, 54–55
poaching, 17
policy analysis, improving data, 43
policy design for environmental health, 63
policy reforms, 46
poor, 10, 15tn, 60, 63
  biomass density increase, 22f
  environmental income, 12–15, 61
  participation in environmental
    management, 65, 66
  resource dependency, 65
population growth, 7
population in poverty, 1
poverty, 22–24
  and environment degradation link,
    11–12, 23–24
  and PES, 55–56
  and resource availability, 20
  and resources and U-5 mortality, 2t
  Cambodia, 12b–13b
  environment linkages, 5f, 8–9
poverty indicators, 2–3, 4–5, 4f
poverty reduction, 1–2, 11, 24
  and environmental management,
    9, 25, 64–65
  sustainable resource use, 65–66
poverty traps, 7, 62
power-sharing agreements, 55

privatization of common property, 16
privatization of water, 47, 51b
property rights, 14, 46, 53–54
  common, 47–50, 52–53
public programs, 36b, 42

R

randomized experiments, 59
reform, 46, 60, 64–65
  effects on other sectors, 8
  originating sector, 45
Regional Integrated Silvopastoral Ecosystem
    Management Project, 56
research problems, 31–32
resources. See natural resources
respiratory illness, 39, 40, 44n, 63
rich, 15tn

S

sample selection bias, 32
sanitation access, 3t, 38, 63
  child mortality, 35, 38b, 42
  diarrheal morbidity, 35
  latrines, 44n
selection bias, 48b
shocks, 6, 16, 62
silvopastoral practice, 56
stove programs, 44n
stunting, 3t, 38

T

Tanzania, 23
Tawahka community, 17b
timber, 22
time-series data, 59

U

under-5 (U-5) mortality, 2t, 3, 3t, 32, 63
  electricity access, 34
  water access, 33–34
underweight, 38

V

variable bias, 32

W

water, 31, 63
water access, 3t, 33, 38, 57
  child mortality, 34, 38b, 42
  diarrheal morbidity, 35
  piped and impact on child diarrhea, 36

water privatization, 47, 51*b*
water purification, 35, 36*b*
water-hygiene–infectious agent
    transmission, 34
weight for age, 38
welfare. *See* household welfare
welfare index, 22*f*
welfare indicators, 52
wildlife poaching, 17

wildlife, legal access to,
    47, 48–49
women, 37, 40
World Bank, 1

Z

Zanzibar, 23
Zimbabwe, 14

ECO-AUDIT
Environmental Benefits Statement

The World Bank is committed to preserving endangered forests and natural resources. The Office of the Publisher has chosen to print *Poverty and the Environment: Understanding Linkages at the Household Level* on recycled paper with 100 percent postconsumer fiber in accordance with the recommended standards for paper usage set by the Green Press Initiative, a nonprofit program supporting publishers in using fiber that is not sourced from endangered forests. For more information, visit www.greenpressinitiative.org.

Saved:
- 14 trees
- 10 million Btu of total energy
- 1,210 lb. of CO_2 equivalent greenhouse gases
- 5,023 gal. of waste water
- 645 lb. of solid waste